Advanced Lean Thinking

Proven Methods to Reduce Waste and
Improve Quality in Health Care

Foreword by
Gary S. Kaplan, M.D.,
Virginia Mason Health System

Joint Commission
Resources

Joint Commission Resources Mission

The mission of Joint Commission Resources is to continuously improve the safety and quality of care in the United States and in the international community through the provision of education and consultation services and international accreditation.

Joint Commission Resources educational programs and publications support, but are separate from, the accreditation activities of The Joint Commission. Attendees at Joint Commission Resources educational programs and purchasers of Joint Commission Resources publications receive no special consideration or treatment in, or confidential information about, the accreditation process.

Senior Editor:
Helen M. Fry

Contributing Editor:
Lori Meek Schuldt

Project Managers:
Andrew Bernotas, Bridget Chambers

Manager, Publications:
Victoria Gaudette

Associate Director, Production:
Johanna Harris

Associate Director, Editorial Development:
Diane Bell

Executive Director:
Catherine Chopp Hinckley, Ph.D.

Vice President, Learning:
Charles Macfarlane, F.A.C.H.E.

Joint Commission/JCR Reviewers:
Patricia Adamski, Diane Kucic,
Louise Kuhny, Leslie LaBelle, Deborah Nadzam,
Klaus Nether, Frank Zibrat

Joint Commission Resources, Inc. (JCR), a not-for-profit affiliate of the Joint Commission on Accreditation of Healthcare Organizations (The Joint Commission), has been designated by The Joint Commission to publish publications and multimedia products. JCR reproduces and distributes these materials under license from The Joint Commission.

Printed in the U.S.A. 5 4 3 2 1

ISBN: 978-1-59940-228-4
Library of Congress Control Number: 2008938825

Requests for permission to make copies of any part of this work should be mailed to

Permissions Editor
Department of Publications
Joint Commission Resources
One Renaissance Boulevard
Oakbrook Terrace, Illinois 60181
permissions@jcrinc.com

FOR MORE INFORMATION ABOUT JOINT COMMISSION RESOURCES, PLEASE VISIT **HTTP://WWW.JCRINC.COM.**

TABLE OF CONTENTS

FOREWORD

The health care industry is under closer scrutiny than ever before in its history.

Why? Because we aren't meeting the expectations of the public or those paying for the services, including the government and businesses. Our customers are dissatisfied because we aren't able to deliver consistently reliable performance and outcomes; because our charges are high relative to the perception of value received; because the public is expressing unprecedented and justifiable expectations for safer care; and because medical workers' morale has reached new lows due to feeling they are unable to do their best work for their patients.

In addition, many companies are finding that the expense of providing health insurance coverage for their workers is no longer sustainable without risking their ability to be competitive in a global economy. The search for answers to these and other challenges has been illusory, and the result continues to be widespread frustration with many aspects of the U.S. health care system.

Since 2001, we at Virginia Mason Medical Center have been trying to address these issues with a new and radically different approach to our work. Teams of physicians, nurses, technicians, managers, and front-line staff report weekly on the *kaizen* activities of the past week and the improvements they were able to test and implement in just a few days. The improvements are made in the ambulatory care environment and the hospital environment. They impact primary care, specialty care, and critical care. Most important, they lead to fewer defects, safer care, and higher-quality evidenced-based care for patients every day.

What these improvements all have in common is that they are identified and implemented by front-line staff who know their work best, utilizing the methods and tools of Lean—what we call the Virginia Mason Production System, based on the Toyota Production System. Our staff are discovering what is truly possible for our patients and, in so doing, are creating a less burdensome, waste-free work environment so they can do their very best for our patients.

Joint Commission Resources (JCR) continues to make a substantial contribution to spreading these methods throughout our industry. The following pages provide the reader with a wealth of knowledge, outlining in detail the tools and methods of Lean and their applications. These tools have been shown to work in every industry in which they have been applied and are now being used successfully in health care. These chapters are invaluable for those who are focused on quality improvement within their organizations and for those who are looking for ways to make care safer and more reliable. Examples from organizations across the country provide practical applications and inspiration.

Unfortunately, learning the tools of Lean is not enough. To truly transform health care and embrace the potential for "zero-defect" care, leaders must also be willing to challenge our old ways of thinking about how we deliver care. We must embrace the power of care delivery teams and not rely on the imperative of the autonomous physician as the key to safer care. We must deliver evidence-based care to every patient whenever possible, and we must develop zero tolerance for the waste that is omnipresent in our processes. Leaders must be willing to take tough stands, challenge the status quo, and take their responsibility as stewards of precious resources seriously.

The time is now for courageous leadership in health care to rise to the challenge. Our patients and our staff are counting on us to lead. Only with strong leadership will we be able to realize the benefits of Lean tools and create a better health care environment for our patients, our people, and our communities. As our *sensei* Chihiro Nakao, a pioneer of Lean manufacturing, has told us, "It is your destiny to be a leader." Readers of this book will find new hope and new ideas and might even be inspired to become the leaders we need to truly keep our promise to our patients.

Gary S. Kaplan, M.D., FACP, FACMPE, FACPE
Chairman and CEO
Virginia Mason Health System
Seattle, Washington

INTRODUCTION

This book is a follow-up to *Doing More with Less,* the 2006 publication of Joint Commission Resources. In order to use the tools in this sequel, a basic familiarity with Lean is required. This edition will provide more in-depth detail for five particular tools: value stream mapping, 5S events, *kaizen* events, error proofing, and Six Sigma. This book also includes seven case studies that provide greater insights into the use of the Lean tools presented here, Lean thinking in general, related Lean tools, and associated quality improvement tools. The last of these case studies also serves as a springboard for implementing Lean in your own health care organization.

Lean and Health Care

Although Lean developed initially in the 1930s as a response to the needs to eliminate waste and improve quality in manufacturing, since the 1980s, Lean thinking (or Lean) has been expanded and adapted to meet the needs of the health care industry. Health care as an industry has experienced significant challenges in providing quality services while enduring great cost pressures. In response, many organizations are looking for improved ways to address these challenges. Lean benefits include the following:

- Valuing diversity by including all stakeholders in problem solving
- Sharing information among employees, which boosts cross-functional understanding and process awareness and in turn decreases rework
- Illuminating where waste occurs in order to promote immediate implementation of solutions
- Paralleling the patient experience
- Giving employees a greater feeling of empowerment and control as they move toward looking clearly at the simple but often overlooked things that impact daily working life
- Fostering team spirit

It is said that using Lean promotes the development of a "community of scientists,"[1,2] whose daily life at work comes to include simple but effective experiments that add to overall improvements over time, in small ways and large. Employees do not wait for reports or directives to look for ways to improve systems and processes. All work becomes highly specific regarding its content, sequence, timing, and outcome.[1] Administrators, once distanced from work on the front line (or *gemba,* in Japanese), come to appreciate what their staff and workers experience. Lean thinking means that all staff share a common goal and a common sense of what an ideal system would be like.

Lean is the evolution of the Toyota Production System (TPS) as well as other quality improvement approaches. As Lean began to move from company to company, it also migrated from industry to industry. Any business in any industry can benefit from the adoption of Lean concepts and methodologies. Although there is no single business or health care organization that is a model for Lean, certain institutions have a clear start and have tracked outstanding benefits from using Lean. Some of these institutions have contributed the case studies that we present in Part 2.

Waste and Variation in Health Care

Lean thinkers and experts are most recognized for their mission to declare war on waste in the workplace. According to the Institute of Medicine, 30 to 40 cents of every dollar spent on health care is for costs associated with overuse, misuse, underuse, duplication, system failure, unnecessary repetition, poor communication, and inefficiency, all of which can be described as "waste."[3]

Waste is defined as anything that does not add value to a product or service from the viewpoint of the customer. In health care, the primary customer is the patient and the patient's family.

More specifically, waste (or *muda*, in Japanese) may occur in time, space, cost, energy, or errors. Any time a process exists, there is potential for waste, or *non-value-added activities.* In a health care setting, for instance, value would include comfort, compassion, competence, and the achievement of desired outcomes. Classifying process steps as either value-added or non-value-added from the patient's perspective is the first step in exposing and eliminating that waste. In Lean thinking, non-value-added work is also subclassified; that is, some of this work may not be considered of value by the patient, but it is required nonetheless. This type of work is referred to as "non-value-added essential."

The proportion of waste in health care has been estimated in a range from 30% to 60%.[4] By one estimate, only about 10% of work performed is considered value-added.[5] Waste can include—but is not limited to—waiting; barriers to flow; handoff breakdowns; errors and mistakes; correcting, revising, or reevaluating; inaccurate information; inaccessible information (for example, when a patient's history is unavailable); shortage or lack of tools or equipment; incorrect or inappropriate equipment; inefficient motions, such an unproductive walking time; unnecessary movement; inaccessible tools or supplies; and inflexible processes (inability to quickly improvise).

A report by the Murphy Leadership Institute[6] posts the following top 10 most wasteful activities in health care organizations:

1. Completing multiple forms for the same task
2. Inefficient shift-to-shift ratios
3. Staff interruptions
4. Hunting for equipment
5. Unavailable or delayed medication administration
6. Long-lasting meetings
7. Searching for or correcting misplaced records
8. Unnecessary or redundant communications
9. Waiting for physician availability
10. Waiting for the delivery of an item from another department

A more inclusive list of wastes is shown in Table 1-1.

Waste arises for any number of reasons. As workers get better at "seeing" the workplace and closely observing the ongoing processes surrounding them, they improve the depth with which they can discern waste. For instance, to the Lean-untrained eye, a busy staff means efficient work. However, if what the staff is doing is not adding value, this "busy-ness" is merely waste masquerading as work.

One huge source of waste is work-arounds, which typically increase the risk for error and reduce the precision of systems that are inherent in quality. A *work-around* is any activity that is implemented to "get around" a problem. Work-arounds may meet a person's immediate needs, but they do not resolve the ambiguities that result in a complex system. People who work at the same tasks, week after week, month after month, and even year after year become used to the work-arounds they have put in place as stopgaps.

TABLE 1-1. Example of Wastes and Variations	
Delays or waiting	Waiting for people (physicians, nurses) Waiting for bed assignments, treatment, equipment, tests, signatures, approval, supplies, or information Specimens waiting in batches for testing Patients waiting due to physician lateness or schedule exceeding capacity Delay in admission to or from emergency department Delays in testing, treatment, or discharge Delay in patient lab test results
Overproduction and overuse	Unnecessary services, such as preparing drug mixtures in anticipation of patient needs Costly tests, such as MRI, CT scans, when not evidence based "Just in case" blood tubes drawn from patients but not used Delays in medical interviews and treatment Medications given to patient/resident early to suit staff schedules Testing ahead of time to suit lab schedule
Motion	Excess movement of people, equipment, paper information, or electronic exchanges; especially the wasteful searching for materials and supplies Excess transport of equipment, specimens, or samples Long walks between departments or areas, for example, from medical clinic to chemotherapy Searching for patients Searching for meds Searching for charts Gathering tools and equipment Gathering supplies Handling paperwork
Extra Processing	Excessive effort, that is, more than is required or requested Ordering more diagnostic tests than needed or required Non-value-added steps in processes Work-arounds Excessive paperwork Repetitive work Using an intravenous line when oral medication would suffice Time/date stamps on labels that are not used Time spent creating a schedule that is not followed Multiple bed moves Multiple testing Retesting Unnecessary procedures

TABLE 1-1. Example of Wastes and Variations, *continued*	
Inventory	More inventory on hand than is required to meet customer needs at moment
	Work piles or Work in Progress (WIP), such as lab specimens awaiting analysis, dictation awaiting transcription, or paperwork in process
	Supplies that are kept on hand and take up space
	Expired test reagents and medications
	Bed assignment and patients in beds
	Pharmacy stock
	Lab supplies
	Samples
	Specimens awaiting analysis
Defects, mistakes, or adverse events	Wrong patient
	Missing information
	Work output that contains errors or lacks something necessary for the next step in the process (for example, incomplete discharge orders)
	Medication errors, including wrong medication and wrong dosage
	Wrong-site surgery
	Poor labeling
	Poor packaging
	Poor storage
	Multiple sticks for blood draws
	Misdiagnosis and diagnostic delay
	Wrong procedure
	Blood redraws
	Poor clinical outcomes (preventable)
	Medication orders not kept up to date after patient is transferred (handoffs)
Transportation	Moving patients for testing
	Moving patients for treatment
	Moving patients
	Moving samples and specimens
	Disorderly schedule for picking up equipment
Human potential[7]	Employee ideas not listened to
	Employees or patients/families not allowed to contribute
	Intimidating employees or professional students[8–13]
	People not acknowledged for contributions and ideas
	Wasted time and faulty organization systems reduced amount of direct patient care time
	Not highest and best use of talent (for example, registered nurses performing nonclinical tasks)
	Low staff job satisfaction, leading to greater turnover rates

Lean Concept
WASTE IN A WORK-AROUND

At one hospital, nurses needed keys to adjust doses of patient-controlled anesthesia (PCA) pumps.[15] Security considerations dictated that each unit would be issued just a few set of keys. As a work-around, nurses spent an inordinate amount time tracking down one of the sets of keys. When a focus was put on this issue and the amount of time was calculated, nurses discovered that on each shift they searched for keys an average of 23 times, wasting 49 minutes in the process. Their solution was to revise the pharmacy mandate: Nurses now each had their own key, which they signed in and out only at the beginning and end of their shift. This practice satisfied the need for security; when it was then deployed throughout the hospital, time spent searching for keys was reduced to almost zero, and time saved totaled 2,900 nurse-hours per year.

Although people have the best intentions, when they are in the midst of their workday and they see a problem, inefficiency, barrier, or irritation, their customary tendency is to perform the work-around rather than do what is necessary to solve the problem. In less-than-ideal environments, the "normalization of deviance" can occur.[14] That is, when individuals ignore or minimize a problem and grow tolerant of things and circumstances that are not quite ideal—or, worse, are ineffective or wasteful—then the "deviant" behavior becomes the norm.

Work-arounds are not tolerated in Lean, and in fact, when converting to Lean, acknowledging when something is waste is perhaps the biggest obstacle to overcome. Instead of forgetting to solve a problem and forming a stopgap measure to work around it, everyone is embraced as a part of seeing and solving. Doing the work is interwoven with the process of seeing it as it is and venturing to do it better not someday, not when it is convenient, but as an integral part of the work itself.

Non-value-added activities increase costs, time, and resources without directly satisfying the customer. Such activities generally show that there is a problem within a process. Often the "pacemaker" process is the most downstream process, but it can be any process that sets the pace for the entire stream. In other words, work flow can go only as fast as its slowest component. The pacemaker process is therefore a critical place for intervention with Lean thinking.

Value-added activities are those activities that change the form, fit, or function of a product or transaction in order to satisfy customers and directly fill a need. Complex processes involved in health care provide never-ending opportunities to eliminate waste and variation. What sets apart companies using Lean thinking is not the solutions they create but their ability to master the process.

Variation is also an enemy of quality and safety. When processes vary, the stage is set for outcomes that also vary. In clinical practice, evidence-based data serve to provide a burning platform, a foundational impetus from which strategies are set.

How Lean Works for Health Care

Although Lean is simple in its precepts, it is not always easy to implement in health care. Barriers include the natural human desire to cling to the familiar, the fear of change, and a lack of understanding of how Lean works. In addition, although Lean health care strives to be patient-centric and to always serve the patient, some aspects of health care may be necessary for delivering care and yet not represent services or products the patient might inherently value. Staff must be won over to Lean thinking to best serve patients as well as other stakeholders. Overcoming this cultural inertia involves hands-on leadership at all levels. Leaders must become teachers and be visible on the front line. As part of a Lean transformation, leadership styles will also require "modernization." The more educated, "psyched," and inspired the staff are led to be, the better the process works.

Lean is a compilation of world-class practices that can improve a health care organization through an evidence-based methodology. Lean is also a management system whose purpose is to eliminate all waste or non-value-added elements by means of daily, hourly, or momentary review. In addition, Lean is about respect for all the people in the organization. Making sure the organization has eliminated waste in its processes sends the message to each worker that "we value you" and "we want to make the work you do as waste free as possible." Lean is not intended to eliminate people but rather to use them more wisely. It is based on reducing costs rather than raising prices or reducing services.

Lean thinking is based on a philosophy that includes the steps of stabilizing, standardizing, and simplifying work processes, regardless of the health care setting:
- *Stabilize:* It is critical that excess movement of work and process variation be removed first. It is difficult to apply Lean when variation exists.
- *Standardize:* Once an area or process has been stabilized, formal standards via work rules, charts, and visual controls can be used to further eliminate variation.
- *Simplify:* As an organization stabilizes and standardizes process, it then can balance the work load among staff to ensure effective and efficient service is provided to the patient.

One of the practices that is integral to Lean and that can help organizations follow the aforementioned steps is the technique of visual aids or controls. This technique uses a visual communication system to control and standardize a process. Examples include checklists, task lists, posted displays of process steps, posted warnings and alerts, story or sign boards, value stream maps, *kanbans* (signal cards), indicators, or color codes. Using these indicators and controls provides the benefit of reducing confusion and stress, encouraging process standardization, and reducing errors.

For example, in 2004, Lean management led to $7.5 million in savings in Park Nicollet Health Services, St. Louis Park, Minnesota.[16] Average patient waiting time at the St. Louis Park Urgent Care facility was reduced from 122 minutes to 52 minutes. The number of phone calls answered within 30 seconds was increased by 560%. The cycle time between when a request for a prescription refill was made to when the pharmacy received authorization was reduced by 79%. The number of medications prepared but not needed was reduced by 30%. After analyzing the variation in surgical instrument preferences and the agreement on standardization of instruments, 40,000 fewer surgical instruments were processed each month.

The Case for Lean in Health Care

Lean practices help a health care organization run as a successful business, focusing on achieving a positive return on investment (ROI) by eliminating, or at least minimizing, non-value-added activities. Such practices might include the following[17]:
- Improving employee retention
- Reducing adverse events
- Reducing admission rate, resulting in more bed turns and increasing, by means of new and expanded services, the opportunity to fill beds

- Reducing length of stay
- Better leveraging available resources
- Improving employee and physician satisfaction and physician/nurse relations
- Increasing nurses' direct patient care time, which will result in improved patient perception of care

Lean thinking contributes to cost savings, increased quality and safety, and increased patient perception of care and staff satisfaction. Although almost all health care organizations employ some form of quality improvement, health care organizations that use Lean have a competitive advantage as an employer because Lean is built on the participation of all employees, allowing them to contribute their ideas, solutions, and creativity.

Although the advantages of using Lean seem self-evident, persuading the organization to adopt Lean practices requires enrolling everyone into the proposed benefits. (See Chapter 12.)

There are two ways of looking at implementing Lean practices into a health care organization. Some observers say that converting to Lean is so challenging that it requires leadership's review of the entire organization—that beginning with one department may be self-defeating because all departments overlap, and one department will differ so radically that improving overall functionality will never stand up. Others say that those who begin too big may end up with a lack of focus and are doomed to fail. The key is to implement Lean while continuing to emphasize current priorities, in the process adjusting focuses and projects over time. As Lean thinking and projects increase, organizations can continue to assess, reassess, and plan for including more and more of the organization into Lean.

To be the advocate or ambassador, or eventually the bona fide champion, for Lean takes commitment on an individual as well as team level. But as Lean experts and those well experienced in its benefits will testify, Lean has individual and personal benefits, too. Personal spaces begin to shine and show order, and in these settings, productivity and satisfaction increase. Creative signals and visual controls are used to lighten, brighten, and clarify. Searching and hunting decrease. Time is freed up for what is valuable to the patient, the staff member, and the organization as a whole. Those who take on Lean fully and passionately see many benefits. As you get more familiar with Lean and its tools, discuss them with others in small and large forums to hear and share your stories. And be sure to have fun!

About This Book

Although we focus on just five tools in this book, Lean is not simply a toolbox but an overall organization perspective. It is true that some organizations "cherry-pick" their Lean tools, but adhering to a total Lean management philosophy means implementing Lean steps throughout the organization as its overall strategy for quality and safety. Selecting Lean tools first and then as an afterthought deciding what you need to accomplish with them is putting the cart before the horse.

A Lean culture must be the foundation upon which the tools are used and the benefits accumulated. As you proceed with Lean thinking and working, search for ways to change your organization's culture in addition to seeking any quick, measurable results. Leaders must be informed, vocal champions for "going Lean." A changed culture assures success. All involved, from staff to vendors to top organization management, are allies in this crusade.

Part 1: The Lean Toolbox

Each of the chapters in Part 1 is organized as an easy-to-read guide. The *At a Glance* feature presents the tool description, purpose, responsible staff members, duration, and process steps. A number of related Lean and other quality improvement tools have been integrated throughout the book and highlighted as "tool tutors" to provide examples and explanations. These related tools and concepts, which operate hand in glove with the five Lean tools detailed in Part 1, appear in boxes labeled as follows:

- *Tool Connections* explain related Lean and performance improvement tools.
- *Lean Concepts* define foundational concepts for thinking Lean.
- *Tool Tangents* highlight short examples of a Lean tool or concept in action.

The following five chapters appear in Part 1:

- Chapter 1: Value Stream Mapping
- Chapter 2: 5S Event
- Chapter 3: *Kaizen* Event
- Chapter 4: Error Proofing
- Chapter 5: Six Sigma

Part 2: Lean Applications

Part 2 highlights seven case studies of health care organizations that have begun to implement Lean. They provide insight into how Lean tools are used or adapted to varying settings. The *Case At a Glance* feature includes the organization, Lean project, tools used, and primary outcome. Those features are described in further detail in the study itself. The case studies in Part 2 highlight the following Lean tools and settings:

- Various Lean tools in the hospital setting
- *Kaizen* in two ambulatory care settings
- *Kaizen* in the home health care setting
- A3 in the behavioral health care setting
- Process mapping in the laboratory setting
- Use of key leadership concepts to integrate Lean into the health care culture

The following chart summarizes which tools are used in which case studies.

Tool/Concept	Chapter Case Studies
Value stream mapping	6, 7, 8, 9
5S	8, 9
Kaizen events	6, 8, 9, 10, 12
Error proofing	7
Six Sigma	7, 12

Chapter 12, "How to Make Lean Work for Your Organization," provides a review of one organization's overall foray into Lean thinking. Cancer Treatment Centers of America is one of only a handful of organizations in the United States that have taken Lean on as an institutional foundation and have based their strategic plan around Lean philosophies and methods.

A reference list at the end of the book identifies all the resources used in preparing this book. Finally, a glossary includes definitions of Lean and related tools as well as other concepts that are integral to Lean management.

Terms Used Throughout

Lean operates with one clear focus: the customer. Depending on the health care setting, the customer might alternately be called a patient, client, or resident. In most cases we have used the term *patient* to serve overall. Although Lean's operating principles, specifically as they are rooted in the TPS, have Japanese roots, we have included only the Japanese terms that are used frequently in Lean.

Acknowledgments

We would like to thank the individuals, institutions, and organizations that contributed information for this book. We are especially grateful to the many people who contributed their time and information to construct the case studies provided in Part 2 as well as the project example in Chapter 5, "Six Sigma." They are listed here:

- Tricia Brown, Vice President, Corporate Affairs; Irene Magee, Vice President and Director of the Medical Equipment Division; Northeast Health, Troy, New York
- Herb De Barba, Vice President for Lean Six Sigma at Cancer Treatment Centers of America, Midwestern Regional Medical Center, Zion, Illinois
- Trent Gordon, F.A.C.H.E., Business Development and Planning; Don Calcagno, Vice President of Performance Improvement; Advocate Good Shepherd Hospital, Barrington, Illinois
- Pam Guler, M.H.A., Administrative Fellow, Six Sigma Black Belt; Lisa Johnson, M.S., M.S.N., R.N.-C.N.A.A., CNO, Vice President Patient Services; Morton Plant Mease Health Care, Clearwater, Florida
- Wayne Ketron, Executive Director Performance Improvement and Safety, Six Sigma Master Black Belt, Wellmont Health System, Kingsport, Tennessee; Terry Eads, Director Quality Accreditation and Risk Management, Wellmont Bristol Regional Medical Center, Bristol, Tennessee
- Sandra Kiesel, Principal, Go Lean, Lake Orion, Michigan
- Paul Melinkovich, M.D.; Patricia Gabow, CEO; Denver Community Health Services, Denver, Colorado
- Bev Moncy, Senior Consulting Manager; Tom McFadden, Consulting; Karie Coonrod, Consultant; GE Healthcare
- Jan Pringle, Ph.D., Director, Program Evaluation and Research Unit, School of Pharmacy, University of Pittsburgh
- Charleen Tachibana, R.N., Senior Vice President, Hospital Administrator, and Chief Nursing Officer; Cathie Furman, R.N., M.H.A., Vice President of Quality and Compliance; Alisha Mark, Media Relations Manager; Virginia Mason Medical Center, Seattle, Washington

REFERENCES

1. Spear S.J.: Decoding the DNA of the Toyota Production System. *Harv Bus Rev* 77(5):97–106, Sep.–Oct. 1999.

2. Bushell S., Shelest B.: Discovering Lean thinking at Progressive Healthcare. *Journal for Quality & Participation* 25(2):20–25, 2002.

3. O'Leary D.S.: Reducing waste and improving health care quality and safety in organized health care settings. Presentation at Joint Commission Conference on Waste and Inefficiency in Healthcare, Chicago, Mar. 27–28, 2008.

4. Hadfield D., Holmes S.: *The Lean Healthcare Pocket Guide: Tools for the Elimination of Waste in Hospitals, Clinics and Other Healthcare Facilities.* Chelsea, MI: MCS Media, 2006.

5. Zidel T.G.: A Lean toolbox: Using Lean principles and techniques in healthcare. *J Healthc Qual* 28(1):W1-7–W1-15, 2006.

6. Lazarus I.R., Andell J.: Providers, payers and IT suppliers learn it pays to get "Lean." *Managed Healthcare Executive*, Feb. 2006.

7. Graban M.: *Lean Healthcare.* LeanBlog.org, Jan. 2008. http://www.superfactory.com/document-archive/newsletter-articles/73-lean-healthcare.html (accessed Mar. 21, 2008).

8. Hagstrom A.M.: Perceived barriers to implementation of a successful sharps safety program. *AORN J* 83(2):391–397, 2006.

9. Institute for Safe Medication Practice (ISMP): Intimidation: practitioners speak up about this unresolved problem (Part 1). *ISMP Medication Safety Alert! Acute Care.* Mar. 11, 2004. http://www.ismp.org/newsletters/acutecare/articles/20040311_2.asp (accessed May 8, 2008).

10. Institute for Safe Medication Practice (ISMP): Intimidation: Mapping a plan for cultural change in healthcare (Part 2). *ISMP Medication Safety Alert! Acute Care.* Mar. 25, 2004. http://www.ismp.org/Newsletters/acutecare/articles/20040325.asp (accessed May 8, 2008).

11. Seiden S.C., Galvan C., Lamm R.: Role of medical students in preventing patient harm and enhancing patient safety. *Qual Saf Health Care* 15(4):272–276, 2006.

12. Weinberg N.S., Stason W.B.: Managing quality in hospital practice. *Int J Qual Health Care* 10(4):295–302, 1998.

13. Wood D.F.: Bullying and harassment in medical schools. *BMJ* 333(7570):664–665, 2006.

14. Groom R.: Normalization of deviance: Rocket science 101. *J Extra Corpor Technol* 38(3):201–202, 2006.

15. Spear S.J.: Fixing health care from the inside, today. *Harv Bus Rev* 83(9):78–91, 158, 2005.

16. Abelson D.: Lean production efforts help save $7.5M in 1 year. *Healthcare Benchmarks Qual Improv* 12:137–138, Dec. 2005.

17. Burnes Bolton L. Defining and achieving efficiency: Improving RN time in direct patient care. Presentation at Joint Commission Conference on Waste and Inefficiency in Healthcare, Chicago, Mar. 28, 2008.

PART ONE

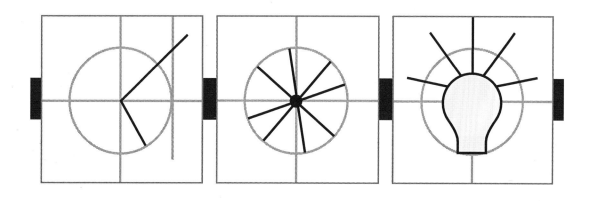

THE LEAN TOOLBOX

CHAPTER ONE

VALUE STREAM MAPPING

Value Stream Mapping At a Glance

TOOL DESCRIPTION	Value Stream Mapping is a tool for helping to observe and come to understand the flow of material and information along the value stream. The value stream includes everything that must occur in order to provide a service to a patient.
TOOL PURPOSE	Allow the team to "see" the work flow and information required for a given process(es) and linked by a common theme (for example, following a document through to its end destination or following a patient through a health care procedure):
	• Clearly identify waste
	• Create a common vision for everyone connected to the value stream
	• Provide a road map to allocate appropriate resources
WHO IS RESPONSIBLE?	A cross-functional team representing the value streams.
HOW LONG DOES IT TAKE?	It depends on the complexity of the process. Frequently, it takes 1 to 3 days to create a current state map and a first attempt at creating a future state map.
STEPS IN THE PROCESS	1. Create a current state map.
	2. Create a future state map.

The value inherent in any organization is the end result of a long sequence of steps: the value stream. The *value stream* involves everything that must happen in order for a patient to receive a desired health care product or service. To end up with the proper outcomes, all steps must be conducted in the proper sequence at the appropriate time. Delivering the value stream to the patient means crossing boundaries of departments to incorporate the total system.

Health care is often organized around functions. In a hospital, for example, issuing medication is the function of the pharmacist, and drawing blood is the function of the phlebotomist. Separating activities by function can fragment care. Ambiguity over who is responsible for what task can lead to inefficiency and, in some cases, catastrophes. Communication breakdowns between the various "silos" of health care (for example, between pharmacy and nursing or between managers and suppliers) can create risks for the patient.

Value is determined based on benefit to the customer, and in health care, the major customer is the patient. Focusing on that patient can help integrate the individual elements of care and the various roles delegated in care into a safe and effective whole. Value stream mapping is a way to "see" the various elements of activities and any barriers that exist across departments and units. Using value stream mapping helps workers focus on ultimate use for the benefit of the patient and integrate activities across departments.

The value streams of an organization are those processes that support or produce the core products and services. In health care, a core product or service can be an office visit, a hospital stay, a diagnostic or laboratory test, or a call to an office to get a prescription refilled. In value stream mapping, each process step for each product or service is identified. Value stream maps are intended to capture the essence of the value as the patient experiences the process. As viewed by Toyota, all learning comes from work on the production floor (in Japanese, the *gemba,* or front line). In health care, the "unit of production" is the patient. Value stream maps should not be confused with process maps—diagrams that are intended to outline the process with precision.

Learning to differentiate what is value and what is waste is a skill. Value stream mapping allows the process disconnects and barriers to flow to stand out. Lean thinking involves learning to "see" with a meticulous attention to detail so that the scene or process being studied begins to speak to the researcher. Using the value stream mapping tool is a first step in what the Japanese refer to as "putting on your *muda* glasses" (*muda* being the Japanese word for "waste") and endorsing *genchi gembutsu,* translated as "go see for yourself."[1]

Steps in the Process

The value stream is a process, and the value stream map is a picture of everything that needs to happen in order to supply a product or service to the patient. Although a value stream may cross boundaries and overlap with other value streams, it is important to focus. In general, every area of strategic flow in each unit or department should be mapped separately. From doing so, a graphical representation can be made of existing constraints and defects.

There are two overall steps to value stream mapping: (1) creating a current state map to identify the waste and what is not being done in this area and (2) creating a future state map that serves as the destination of where you want to go, that is, how the process should flow.

The steps to create a value stream current state map include the following:

1. Form the cross-functional team.
2. Schedule a walk-through and explain its purpose to unit, section, or department staff.
3. Identify mapping tools and prepare supplies (what you'll need).
4. Execute the walk-through to observe value and its absence: Ask questions and discuss with staff of the unit, section, or department.
5. Create the map.

Tool Tangent

IDENTIFYING WASTE

The following acrostic is designed to help you see and remember eight common wastes in work and processes.

D **Defects**
Defective products or information leads to waste̵ for example, missing/incomplete information, medication errors.

O **Overproduction**
Producing more products or information than needed is wasteful (the most significant form of waste)̵ for example, any service that does not help heal the patient.

W **Waiting**
People, material, or equipment not being acted upon

N **Not highest and best use of talent**
Correct use of people and ideas

T **Transportation**
Moving patients, meds, etc.

I **Inventories**
Material or information that sits can potentially become lost or get damaged̵ for example, bed assignment without patients.

M **Motion**
Any motion that does not add value̵ for example, searching for patients, meds, charts, etc.

E **Extra processing**
That which is not essential to value added to customer̵ for example, typing and writing the same information

Source: Sugiyama T.: *The Improvement Book.* Cambridge, MA: Productivity Press, 1989.

Lean Concept

FLOW

Flow is the continuous movement and progressive achievement of tasks along a value stream. To accomplish better flow, eliminate constraints or barriers that impede movement of the patient or product from one operation or process to the next.

Continuous flow designates the state of zero waste and no holdups in pulled time. In other words, things flow as they should, when they should, how they should.

The flow of information throughout health care organizations represents a major opportunity for waste and interrupted flow. Creating a communications diagram can help staff visualize the flow of information and identify opportunities for streamlining information flow to ensure that the right staff receive the right information at the right time. The purpose of such a communications assessment is to help see the movement of information and make overproduction of information obvious. Figure 1-1, which follows, provides a sample before-and-after communications diagram.

FIGURE 1-1. Communications Flow

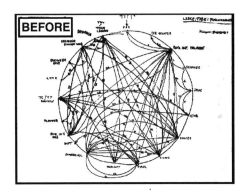 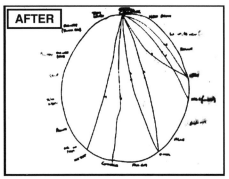

A communications diagram similar to the one pictured here can help a team more readily identify disruptions and waste in the flow of information for a given process. Take a look at your "before" diagram and keep asking, "Why?" Why does the information need to change hands? A very different "after" diagram is likely possible.

Source: Advocate Good Shepherd, reproduced by permission.

Once the current state map is created, continue to discuss, brainstorm, and communicate with staff. You will then create in iterations the future state map.

With a map of the future state, and comparing it with the current state map, the team may identify the gap between current state and future state, develop a plan to close the gap, outline the steps of implementation in the plan to remove the waste, and implement change. The team should continue to gather data and sustain change, reporting progress, pitfalls, and prognosis.

Within any health care system, practices are interdependent and are implemented across multiple tasks and activities. For instance, managing the "pull" system of a patient's journey through a clinic depends on the timing and performance of each preceding activity. All aspects (timing, effort, waste, and so on) must be measured so that the team can specify and synchronize tasks

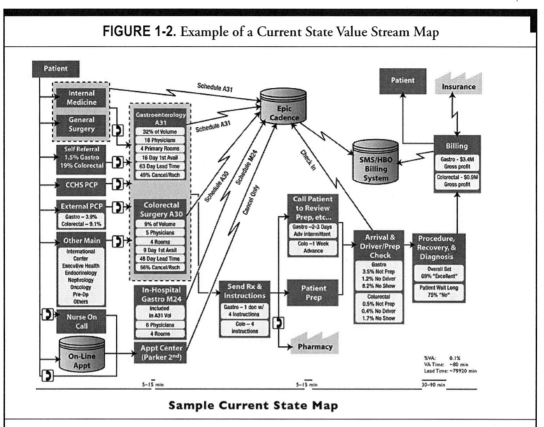

FIGURE 1-2. Example of a Current State Value Stream Map

Sample Current State Map

The goal of mapping is to create a visual representation of every process in the patient and information flows. This current state map was created as a first step toward improving the value stream for colonoscopy patient screening.

Source: Wince R.: Improvement in healthcare is possible—Just "be the ball." *Patient Safety & Quality Healthcare,* May–Jun. 2007. http://www.psqh.com/mayjun07/improvement.html (accessed Jul. 7, 2008).

and outcomes. Guessing or speculating does not suffice. Lean processes focus on direct reporting of measurements at the source of activity. Such reporting provides real-time visibility across the production process and forms a basis on which to recognize change and improvement.

The measurement system you choose to use in creating your map must be simple and devoid of too many metrics. The map will measure the process, not the people, and should not overly focus on financial measures. The measurements must be timely (hourly, daily, weekly, monthly) so that corrective action can be taken as soon as possible. Therefore, an important step is to quantify the end-to-end process in terms of time. This estimate includes not only the total time from process beginning to end but also a quantification of the *value-added time* (the time element that the customer is willing to pay for)—which in some processes can be less than 0.5%.[2] (See the section "Time Observations" later in this chapter for more discussion.) The ultimate goal of your subsequent future state map will be to create a value stream that provides resources just in time (see the Lean Concept in Chapter 2). An example of a current state map is presented in Figure 1-2 above.

FLOWCHARTS

Flowcharts show the mapping of flow of the patient or a service for the patient. Basic flowcharting symbols include Supplies, Database, Sequence, *Kanban*, Information, Inventory, and Improvement. Each has its own icon. A flowchart is similar to a value stream map, but it is bigger and much more detailed. Process flowcharting is used to analyze and improve a process, but value stream mapping cuts across process, departmental, and functional boundaries and existing performance measurement systems.

Tool Tangent

VALUE STREAM

At West Penn Hospital in Pittsburgh, the presurgical unit prepared about 40 patients for scheduled surgery. The role of drawing the patient's blood was ambiguous, handled variously by the examining nurse, a technician, or in some cases, not at all. The blood work for one in six patients failed to be completed prior to the patient's going to the operating room (OR), which resulted in the OR staff idling (wasted time) at an estimated cost of $300 per minute. This delay also created a problem for the patient, who was anxious about the procedure and had been fasting since the night before. Such delays bottlenecked the OR for the rest of the day. After a review of the current state map, ambiguities were attacked: Visual indicators were introduced for the patients whose blood had been drawn, a particular staff member was designated to take all blood samples, a visual card was placed on the rack to signal that the patient had been registered, and a closet was converted to a room for drawing blood. The number of patients sent to the OR without blood work declined to zero, and the patient's comfort and dignity were respected and provided for.[3]

Whereas the overall goal is to create a value stream map of an entire facility's processes, attempting to do so all at once is impractical and bound to fail. Start slow. Small changes taken together accomplish big gains. Choose one area and one process flow. By doing so, it will be possible to determine non-value-added operations and to envision what could make things more efficient and effective and what can best deliver the end service to the patient.

Go for the "low-hanging fruit." What is an area in which there are problems but not ones that are overwhelming? Choose something that is doable and achievable in a relatively short time. The value stream represents both actual physical flow and flow of information related to the product or service. For a patient, this flow might include throughput from admission to discharge. For a product, such as a laboratory test, it might include individual steps from arrival to delivery.

It is important to remember that following the flow of value means focusing on a moving target. All systems change. Continuing your Lean observation requires flexibility and tenacity.

You will also, and fundamentally, need buy-in from every member of the team. Team members may include not only people in the unit or department tackling value stream activities but also those at the top of the organization as well as those in departments that are operating overlapping or synchronous functions.

To design value stream maps accurately, all activities and all information must be identified before placing them appropriately on the map, and the patient must be the center of value.

Creating a Current State Map

To create the current state map, it will be important not only to observe the physical area but also to identify communication methods (Web site, e-mail, data tracking and collection, paper reports and memos, and so forth) as well as the silos that contribute to or participate in the process. The ultimate objective is to scrutinize the value stream for waste as seen through the patient's eyes.

A cross-functional team is best composed of representatives from various functional groups in the organization. The team members should be led by middle managers responsible for implementing change and should include staff who have intimate knowledge of the process being studied.

One of the first steps the team may take to create a current state value stream map is to become familiar with the processes involved. The team should begin by creating a basic physical layout of the station, unit, or department. You may wish to create a preliminary, informal "skeleton" map to begin the process, one that employs stick figures and makeshift icons; they can be modified further later. Sidebar 1-1 outlines the resources you'll need to gather to create your process map.

The Walk-Through

Schedule the walk-through at the *gemba* (unit, front line, or production floor) level. Be sure everyone on the team knows that no actions or attributes of the process must be changed at this point. This is a time to notice and measure. Don't estimate and don't work from memory. Walk from one process step to the next, talk to members, and take measurements as needed. Do not divvy up process steps between team members. Everyone needs a clear picture of the entire value stream. Investigators observe the area in real time, changing nothing, capturing everything—including where items, information, or people are stationed, for how long, and where they go, from one step to the next. Do not get concerned about how long tasks take. The walk-through allows the observers an opportunity to gain understanding of the processes and determine the data collection strategy for the next step in developing a value stream map: time observations.

SIDEBAR 1-1.
What You Will Need to Map a Process

1. Patience and good observational skills, including open eyes, ears, and senses; whatever you need to notice constraints and non-value-added steps

2. A stopwatch with a second hand or a digital timepiece

3. Paper (11 x 17 size) and pencil, which is easier than pen for erasing as the activity continues

4. An electronic or paper method for mapping. The organization may want to purchase one of the numerous value stream mapping software products on the market to be used for reporting, but such software is not necessary to perform the activity.

5. A list of value stream mapping icons that is easily seen or distributed

6. An observation sheet for clearly defining the differences between operation, process, work, and task

7. A separate sheet for questions or Post-it® notes in the "Remarks" column

8. A video camera, if possible, with time/date features turned on, to allow observation without interrupting the flow process. Videotaping will also allow others to view the process and comment at a later time, will make rechecking times possible, and will let the observer ask questions in real time and then return to continuous time taking. Make sure to seek institutional and individual permissions in advance for video use.

9. A pedometer, used by some groups to measure distance traveled in steps

---

Ensure that the team reaches consensus about what to collect and how to measure. After the data are collected, ensure that the team agrees on the accuracy of the information on the current state. Finalize the current state map in another location, such as a conference room that is well lit and free of distractions. Finalizing the current state map usually takes 1 to 1.5 days. Keep it simple and confine the entire value stream to no more than one 11 x 17 (A3 or legal-sized) piece of paper.

Tool Connection
A3 PROBLEM SOLVING

A3 is a Lean tool introduced by Sobek and Jimmerson in 2004.[4] This activity is named for the size of A3 paper stock, which is similar to ledger-sized (11 x 17) paper, because this is the approximate size of paper on which these reports must fit. This reporting system is used to document efforts to solve problems concisely and to promote understanding of processes by seeking consensus of all parties that will be affected by proposed changes.

The A3 tool allows all stakeholders, in all areas and departments of the health care setting, to create and implement a problem's solutions. A simple template serves as the basis of a reporting system that doubles to promote understanding of work processes and to gain consensus of all parties that will be affected by proposed changes. The A3 can be completed by staff, helps eliminate non-value-added activities and add value, addresses a specific problem in a systematic manner, identifies and quantifies the current condition, guides the system for solving the problem, and keeps all parties involved and focused along the way. Each A3 should start with a problem statement, worded specifically, such as "Patient care is delayed by 20 minutes waiting for STAT labs to be completed." The statement should be "SMART":

- **S**pecific
- **M**easurable
- **A**ttainable
- **R**elevant
- **T**ime-bounded

Source: Cancer Treatment Centers of America (CTCA): A3: A tool that enables the empowered stakeholder. In *Organizational A3 Textbook.* Zion, IL: CTCA, 2007.

Time Observations

Time is the primary metric in a value stream map. True flow is possible when the process time and lead time are equal. There are two categories of time to measure for a value stream map:

- *Process time:* The time it takes to do the work without interruptions; also known as "touch time," "work time," or cycle time
- *Lead time:* The time that elapses between when work has been made available until it has been completed and passed along to the next step in the process; also known as "throughput," "turnaround," or "elapsed time." Lead time is equal to process time plus waiting time.

Observe the sequence in which the work is performed. If there is no real sequence that is followed and steps are random, work with the person to determine a sequence and request that the sequence be followed during observation. Break a sequence down into individual steps. Concentrate on the work sequence, not the process sequence. Use a separate time observation sheet for each person who performs in the total process.

Once the sequence and process are defined, enter each separate task on the observation form. (A sample time observation sheet is reproduced in Figure 1-3.)

Do not stop time taking after each task in a process. Complete observations of the whole process so as not to interfere with the data. According to Thomas Zidel, author of *A Lean Guide to Transforming Healthcare,* the rule of thumb when it comes to time observations is "the more, the better."[5] The map should show the complete process from "order" to "receipt." Add both value-added and non-value-added elements, but do not add them in the same task. For example, if a person must walk in the process of performing a sequence of work, do not include the walking with the work itself. Split out your time taking as "walk," "get item," and "return"—three different tasks.[6]

Make sure to note time elements as they actually happen, not as you have been told they should happen. Keep timing during interruptions, but record the interruption in a Remarks column. This column provides an area to record reasons for variations in the process sequence: interruptions, unusual occurrences, and such events as waiting or searching (which would be denoted as non-

---

FIGURE 1-3. Time Observation Sheet

	Task	Task Description	1	2	3	4	5	6	7	8	9	10	Best Time	Remarks
	1	Time pulse and record	:58 / 58	4:08 / 43	7:31 / 53	10:57 / 57							43	
	2	Get thermometer	1:06 / 8	4:17 / 9	7:40 / 9	11:05 / 8							8	
	3	Take temperature and record	2:11 / 65	5:26 / 69	8:46 / 66	12:01 / 66							65	
	4	Return thermometer	2:18 / 7	5:34 / 8	8:46 / 10	12:08 / 7							7	
	5	Get blood pressure cuff	2:31 / 13	5:49 / 15	9:00 / 14	13:29 / 81							13	Observation 4 cuff is broken, nurse cannot find spare
	6	Obtain blood pressure and record	3:13 / 42	6:27 / 38	9:39 / 39	14:06 / 37							37	
	7	Return blood pressure cuff	3:25 / 12	6:38 / 11	9:50 / 11	14:18 / 12							11	
	8													
	9													
	10													
	11													
	12													
	13													
	14													
	15													
		Observed cycle time:	3:25	3:13	3:22	4:28							3:04	

Time Observation Sheet

Process observed:
Observe vital signs

Observer:
John Doe

Date:
01-01-2001

This is a sample sheet showing observations of vital signs.

Source: Zidel T.G.: *A Lean Guide to Transforming Healthcare: How to Implement Lean Principles in Hospitals, Medical Offices, Clinics and Other Healthcare Organizations.* Milwaukee: ASQ Quality Press, 2006, p. 56. Reprinted with permission from *A Lean Guide to Transforming Healthcare: How to Implement Lean Principles in Hospitals, Medical Offices, Clinics and Other Healthcare Organizations,* ASQ Quality Press © 2006 American Society for Quality. No further distribution allowed without permission.

value-added activities). To keep remarks clear, link each remark to an observation. For example, if a technician answers the phone during a task, chart "phone call" next to "Observation 6."

Do not cause your own interruptions of the work process while observing processes. If a clarification is needed, jot down your question and ask it after the observation is completed.

Observe at least five cycles before documenting times on the observation sheet. If you can observe more, do it. This will help expose more non-value-added activities. Distractions such as searching for a new blood pressure cuff, a clean syringe, or a proper form are important to record.

Write the time of the first observation in the shaded box associated with the first task. To fill the remaining shaded boxes, subtract the recorded time for the previous operation from the time recorded for the current operation. Enter that value in the corresponding shaded box. Continue to do so until an operation task time is calculated for all observations. Record task times in the shaded boxes using a colored pen. This makes it easier to differentiate between recorded time and task time.

In the column labeled "Best Time," record the lowest task time observed for the related task. To complete the form, add up the times from each observation column and the Best Time column, and enter the number in the Observed Cycle Time boxes on the bottom of the form.

Document tagging, a technique for accurately capturing the amount of time it takes for a chart, document, or patient to travel through a process, area, or value stream, may be useful to reveal cycle times. The document or item is "tagged" with start dates and times, and a time observation is made when the document reaches its interim and ultimate destinations.

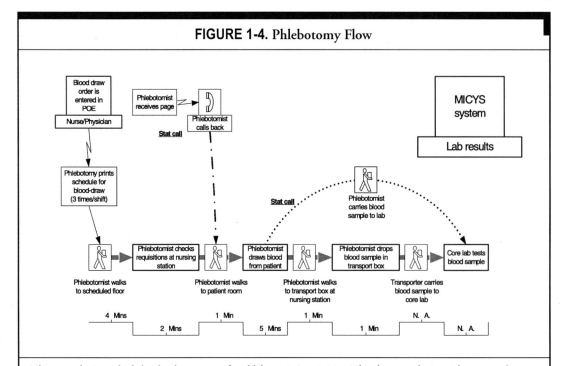

FIGURE 1-4. Phlebotomy Flow

This map depicts a high-level value stream of a phlebotomist's activities. This diagram distinguishes non-value-added activities, such as travel time, from value-added activities, such as the blood draw itself. The time for value-added and non-value-added activities is represented by the "castle wall" line at the bottom of the diagram. The non-value-added activities are depicted through the high portion of the wall, and the value-added activities are represented through the dips in the wall. The non-value-added time may be further distinguished into necessary and unnecessary activities. The necessary non-value-added activities should be minimized and the unnecessary activities eliminated. The phlebotomist's communication activities, from the point of receiving a request for a blood draw to creating the product of a laboratory result, are also shown.[5]

Source: Gabow P., et al.: *A Toolkit for Redesign in Health Care.* Agency for Healthcare Research and Quality (AHRQ). Prepared by Denver Health under Contract No. 290-00-0014-7. AHRQ publication no. 05-0108-EF, Sep. 2005. http://www.ahrq.gov/qual/toolkit/ (accessed May 6, 2008), reproduced by permission.

As you examine each step of the current process, scrutinize the process steps. Consider whether each step in the process produces the desired output, as defined by the patient, and whether a given step produces the same output every time. Ask yourself whether an activity adds value to the process—that is, whether the patient would be willing to pay for this activity. Look at the problems that have been found in the process steps because they will help you determine what your improvement priorities should be. Think about all your findings as you prepare to construct the current state map. Keep in mind that you will need to use the map you create to explain the work system to colleagues and leadership.

Constructing the Actual Map

After completing the observation process, you are ready to construct the current state map. Make sure the map is clear and concise and that when the team sits down to begin the process of creating a future state map, all members of the team easily can see the current state and agree on the functions and operations that have been captured. Look, sketch, and redraw; erase and redo not to sketch the map perfectly but so it clearly represents work flow. Drawing a current state map perfectly is not the point; the subsequent work of improvement is the objective. An example of a current state map is shown in Figure 1-4.

Expect to see a great deal of overlap and feedback between the current state and future state mapping, which is the next step in the process. Each is not a confined exercise. You will be visiting both continually as you use the value stream mapping tool.

It is important that the observers understand the companion terms *pull* and *push*, as well as the concept of *takt* time, to recognize how value stream mapping differs from flowcharting and process maps. Pull and push rely on patient demand, and *takt* time measures the rate of demand.

Creating a Future State Map

The future state map is not how things are working; it depicts how they would ideally work and the way you envision a value stream. In a leaner value stream, steps flow smoothly and barriers and impediments are detected and eliminated. The future map will eliminate non-value-added steps, depict pull mechanisms, and reflect unobstructed flow of a patient or product through the system.

The steps in creating a future state map are to do the following:

1. Develop a work plan that involves all the various action plans and steps necessary to achieve the goals.
2. Identify those who are responsible.
3. Establish a "by when" time frame to accomplish each step.
4. Determine whether any others must be consulted or brought into the process.

Constructing the future state map gives the team the opportunity to focus attention on the barriers to flow. Examples of barriers might include the following[7]:

- Batches and queuing (In contrast to single piece flow, *batching and queuing* is the mass-production of large lots or parts so that waiting time exists before sending the batch to the next operation in the process.)

Lean Concept
PULL AND PUSH

Pull systems are driven by the needs of the downstream lines of production. By specifying value, identifying the value stream, and creating flow, Lean thinking allows pull to take place. The patients pull the product along rather than having the marketplace push it onto them on the organization's timetable. Pull offers more flexibility and accommodates changes in customer demand. In contrast, *push* systems are driven by the output of the preceding lines. This is work that is pushed along regardless of need or request. It involves providing a service or product in anticipation of a need and is often associated with high inventory and the risk of errors or higher error rates. In Lean, *push must be eliminated* and replaced with a pull system to facilitate flow.

Lean Concept
TAKT TIME

Takt time is the rate of customer demand. *Takt*, a term that came into use by the Japanese in the 1930s, is derived from the German word for "beat," "pulse," or "measure." *Takt* time is calculated by dividing available time by customer (patient) demand and represents the number of times a service is performed on a daily basis. For example, if patients need 300 filled medication orders per day and the pharmacy operates 600 minutes per day (assuming operating time and available time are the same), then the *takt* time is 2 minutes. *Takt* time indicates at a glance whether the process is capable of meeting customer demand without running ahead or behind. *Takt* time is most useful to distinguish production when flow is difficult to see.

Tool Tangent
ADJUSTING FLOW

A hospital pharmacy filled its medication orders in batches. After morning physician rounds and order fulfillment, any orders that came in during the day or changes to existing orders accumulated in the computer until the next batch was run. When pharmacists were able to truly see the value stream map and how this process affected the patient who was waiting for orders and changes to be made, the pharmacy in response changed its process to producing and delivering batches every 2 hours as opposed to every 24 hours. This change served a number of purposes, not the least of which was reducing the incidence of missing medications on the hospital floors by 88%.[4] Searching time by nurses for medications decreased by 60% and times of full stock depletion fell by 85%. But not all solutions work for all settings. When a sister hospital in the same system mastered the process that the original hospital used, its staff uncovered different problems and found different solutions.

- Long setup times (Setup is the time it takes, mentally and physically, to prepare for beginning a particular type of task.)
- Equipment availability and down time
- Excessive approvals
- Psychological issues (avoidance)
- Nonstandard priorities
- Schedule changes
- Shared resources/problems with staff accessibility
- Skill level needs
- Interruptions
- Staffing levels and mix/uneven work
- Physical layout

Many of these barriers result from excessive waits for information, material, equipment, or other people—in other words, waste. For example, in a 1,500-employee organization, if each employee spends even 10 minutes of his or her day in excessive searching or waiting, this amounts to 41.7 hours of wasted time per year per employee.[7] At an average pay of $22 per hour per employee, that amounts to a whopping annual cost of $1,376,100. Improving flow, if even one or two minutes can be shaved from each employee's wasted time, will make a big difference overall.

Approvals and authorizations are another area rich for waste. Where possible, move the responsibility for providing authorization to the lowest level of authority and to those who are closest to the work, but *only* within licensure regulations. Create standard work for decision-making procedures so that time can be measured and controlled.

Lower-level staff newly designated to give approval can be trained, and they should be audited periodically until they can be fully trusted in the responsibility. Another way to shave time and reduce waste in wait times is to design processes that require only a maximum of two signatures. Remember, however, that decreasing waiting times for approval is not an area that should be treated lightly. For example, to help reduce medication errors, a hospital instituted a policy in which two nurses must confer and sign off on distinguishing the adult and pediatric versions of heparin as well as the different dosages of these agents. Requiring sign-off from two nurses may be a solution to protect patients. Don't create shortcuts when value and safety may be jeopardized, and consider that error creates more waste and increases cost.

Brainstorm to create a future state map. In general, there are a few ways to best conduct a brainstorming session. The first way is to invite any participant to call out an idea. But sometimes, those who are shier about speaking out may not share their ideas and those who shine in group speaking situations may overly grab for the spotlight. Writing ideas anonymously on index cards can be a more democratic way to handle that problem. Another approach is to gather ideas through a round-robin process where the facilitator moves around the circle and each person provides an idea. The operative guideline is that any person can pass on a turn if

Tool Connection
STANDARD WORK

Standard work (SW) reduces process steps that are wasteful and creates a plan for executing those steps in a specific sequence. Standard work must be performed according to a rigid script and allows for work load balancing. Standard work should include *takt* time, the work sequence of specific tasks for each individual involved, quality checks, safety precautions, and the minimum inventory of parts or supplies on hand to conduct the activity. A standard work sheet is a visual representation displaying the sequence, process layout, and work units for a process. Using SW for repeated tasks improves quality because everyone is precisely following the same work steps and sequence. The seven steps of this tool are the following[8]:

1. Document the reality: "what is."
2. Identify the waste.
3. Plan countermeasures.
4. Implement changes.
5. Verify changes.
6. Quantify changes (measure savings in cost, time, or effort).
7. Standardize changes.

Standard work "bundles" have been used successfully in bringing to zero the variations around clinical benchmark areas such as those of acute myocardial infarction (heart attack), congestive heart failure, stroke, central line care, ventilator-associated pneumonia, and surgical site infection. SW is also the tactic used in medication reconciliation and with patient identification. In the hospital setting, SW improves quality, reduces safety hazards, and allows housestaff to creatively devise better patient care. Types of SW include process steps with visuals such as laminated "cheat sheets" and pocket guides, checklists, decision matrix flowcharts, core competency lists, process flowcharts, metrics-based process maps, and interdepartmental service standards.[7]

Figure 1-5 shows an example of standard work from one health care organization.

FIGURE 1-5. Standard Work Sheet: Decontamination

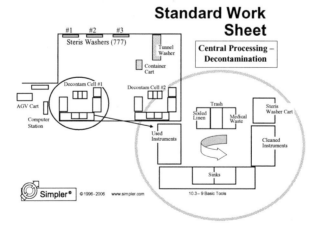

This illustration shows that the logical flow of supplies (in this case, from dirty to clean) can make an individual's work life much easier. Much like a surgeon never having to move during a surgery to access supplies, this person's work station allows him or her to stand in one place and pivot to decontaminate supplies without having to move the feet, thus creating a much more efficient operation.

Source: Advocate Good Shepherd, reproduced by permission.

Tool Connection
BRAINSTORMING

Brainstorming is a process used to elicit a high volume of ideas from a group of people who are encouraged to use their collective thinking power to generate ideas in a relatively short period of time, in an environment that is free of criticism and judgment. Guidelines for successful brainstorming include the following:

1. Never edit or judge anyone's ideas. This is the surest way to quash participation and limit contributions.

2. Encourage participants to contribute all ideas, even ones they believe are "over the top."

3. If possible, ask participants to express an idea in three words or less in order to expedite the process by allowing the facilitator sufficient time to write all ideas on a flip chart or whiteboard.

4. Use color coding, perhaps alternating colors for each idea and numbering them so that they can be clearly differentiated as the process continues.

5. List reduction involves reducing the number of ideas. Do not cross out anyone's ideas; simply circle them with a colored marker. After circling, ideas can be categorized and prioritized.

Lean Concept
FIFO

FIFO stands for "first in, first out" and refers to lanes of production. This tool ensures that the oldest work upstream (first in) is the first to be processed downstream (first out). This could be a raised flag or an e-mail alert. An example of a FIFO system is the supermarket, where foods are stocked in such a way as to allow the consumer to grab the item with the soonest expiration date first or a pharmacy where drugs are stocked in such a way as to allow the pharmacist to grab the oldest-dated drug first.

he or she wishes. The facilitator keeps up the process until everyone runs out of ideas.

Implementing the Future State

To implement the future state, Lean improvement steps must be followed. The actions needed should be scheduled at some point in the mapping process. If they are left to schedule later, later may never come.

Creating a display of the work plan helps move action forward. This display can be any visual that can be updated to show progress, such as a poster, a "thermometer" (as is used in fund-raising), a calendar, or a bulletin board.

Do not travel too far into the future when planning your future state map. Set a target of 3 to 6 months out. Implementing the future state is not an easy task and will take some time.

To signify that patients and products in the future state are being pulled through the system, as opposed to pushed, the map should contain very few triangle (inventory) icons. For patients, there should be little to no waiting; with products and supplies, there is no inventory buildup. Patient waiting time in an ideal future state map has been reduced to zero. Attention also can be given to the concept of FIFO (first in, first out) to ensure that people or products are aligning with flow.

The following design considerations should be incorporated into the future state objective[7]:

1. Improve output quality.
2. Reduce handoffs/merge steps.
3. Stop performing nonessential tasks.
4. Create parallel paths.
5. Implement pull systems.
6. Reduce or eliminate batches.
7. Create standard work.
8. Implement visual management.
9. Eliminate unnecessary approvals/authorizations.
10. Colocate functions based on flow, create cross-functional teams.
11. Balance work to meet *takt* time requirements.

FIGURE 1-6. Prioritization Matrix (PACE chart)

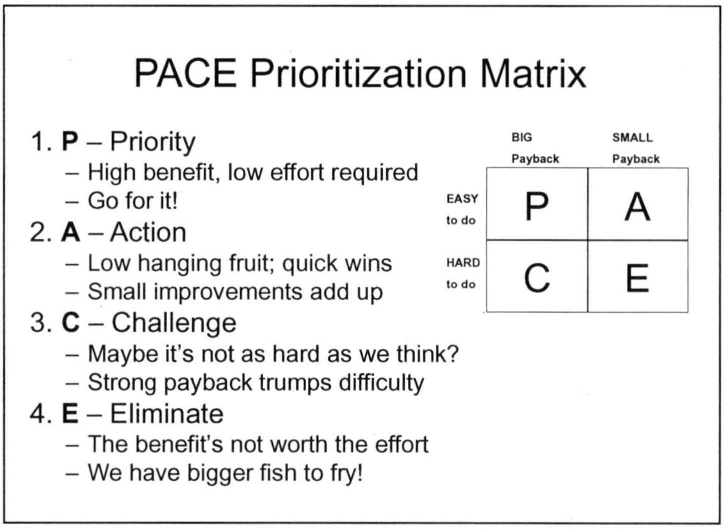

PACE Prioritization Matrix

1. **P** – Priority
   - High benefit, low effort required
   - Go for it!
2. **A** – Action
   - Low hanging fruit; quick wins
   - Small improvements add up
3. **C** – Challenge
   - Maybe it's not as hard as we think?
   - Strong payback trumps difficulty
4. **E** – Eliminate
   - The benefit's not worth the effort
   - We have bigger fish to fry!

	BIG Payback	SMALL Payback
EASY to do	P	A
HARD to do	C	E

The PACE chart can be used to illuminate priorities.

Source: Martin K.: Slide presentation given at Joint Commission Resources Ambulatory Care Conference, Chicago, Oct. 1, 2007, reproduced by permission.

A great map is useless without follow-through. Use the PACE prioritization matrix, as shown in Figure 1-6, to discern what will be easier to tackle and what will take more time, ingenuity, and effort.

To create a future state map of value and aligned with Lean thinking, always draw the patient first and consider improvement from that perspective. Other important points to remember are that cross-training is often key to accommodate isolated silos and separated functions. Remember, future state maps often evolve through many iterations and may become more detailed and finished over months to a year or more. But don't let the big goals obscure the small accomplishments.

An example of a future state map is presented in Figure 1-7. The map is complete when it captures the process from start to finish, is in current sequence, and includes all process steps. Remember to establish a plan for sustainability, which includes documenting standard work processes, designing graphics to communicate and share progress, continuing to collect data and measure *takt* time, assigning project ownership, and transferring ownership with any staff turnover.

Relevant Health Care Applications

Health care organizations have successfully employed value stream mapping to drive Lean improvement events focused on identifying weaknesses in the medication reconciliation process, reducing lab turnaround times, improving insurance claim processing and accounts receivable, tracking the flow of patients or supplies through an area, improving appointment scheduling,

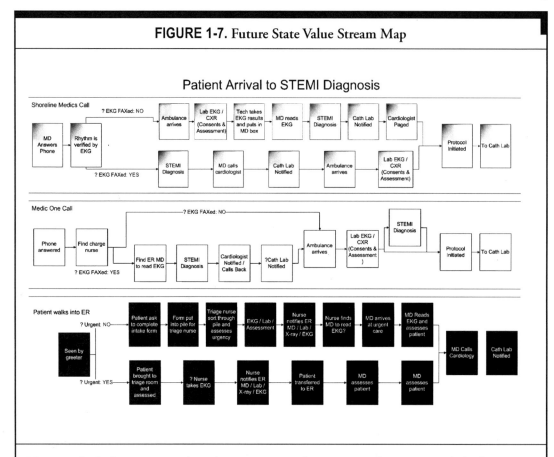

FIGURE 1-7. Future State Value Stream Map

Patient Arrival to STEMI Diagnosis

This example of a future state map shows the patient arriving for ST-segment elevation myocardial infarction (STEMI) diagnosis. MD, physician; ER, emergency room.

Source: Nancy West: Using Lean principles for clinical quality improvement. Slide presentation given as a Qualis Health distant-learning program in a Training Exchange through the AIDS Education and Training Centers National Resource Center, Newark, NJ, Jan. 23, 2008, reproduced by permission.

and visually identifying opportunities in nearly any clinical, financial, or administrative area to reduce effort, time, space, cost, and mistakes.

In the Denver Health system, leaders identified five strategic organization components that were critical to the survival of the organization and then developed value stream maps with specific beginning and end points[9]:

1. Access to care (begins with the patient attempting to access the system and ends with the patient obtaining an appointment)
2. Billing (begins with the generation of a charge and ends with receipt of payment)
3. Outpatient flow (begins with the patient's entry into ambulatory care and ends with the patient leaving the clinic)
4. Inpatient flow (begins with patient entrance into hospital-based care and ends with patient discharge)

5. Operating room flow (begins with the decision for surgery and ends with the discharge of the patient from the postanesthesia care unit)

By specifying and diagramming these maps, the organization worked to modify systems so that employees did not perform tasks having no value, patients did not endure processes having no value, and resources were used only for ends that provided value.

Intermountain Health System (Salt Lake City[10]) created a laboratory value stream map to drive clinical safety improvements. In the anatomical pathology laboratory at LDS Hospital, problems included slow turnaround times for pathology reports (five days compared with desired two days), increased high turnaround time for the transcription reports to the pathologists (five hours), and, as is unfortunately true in many health care settings, frequent labeling errors. After the implementation of value stream mapping, A3 reports, and process mapping, the two-day turnaround time for pathologists' reports was achieved and the transcription report turnaround time was reduced to one hour. Labeling errors also decreased from an average of three per month to one per month.

Community Medical Center (Missoula, Montana[10]) had a problem with missed medications, which was compounded by other departments' perceptions of low levels of service and responsiveness by the pharmacy. The average order delivery time was found to be 38 minutes, and the process held many ambiguities. After employing value stream mapping, the tool "5 Whys" (see Chapter 4) was used to help conduct a root cause analysis. A3 problem-solving reports were then created. The ultimate outcome was that the average number of orders in the system dropped by 32% along with order-to-delivery times. Pharmacists reported less stress in carrying out their responsibilities as well as a 40% decrease in missed medication notifications, moving from an average of 24 per day to 14 per day.

REFERENCES

1. Lean Enterprise Institute. *Home Page.* www.lean.org (accessed Mar. 7, 2008).

2. Wince R.: Improvement in healthcare is possible—Just "be the ball." *Patient Safety & Quality Healthcare,* May–Jun. 2007. http://www.psqh.com/mayjun07/improvement.html (accessed Jul. 7, 2008).

3. Spear S.J.: Fixing health care from the inside, today. *Harv Bus Rev* 83(9):78–91, 158, 2005.

4. Sobek D., Jimmerson C.: A3 reports: Tool for process improvement. Paper presented at the Industrial Engineering Research Conference Proceedings, Houston, May 16–18, 2004.

5. Gabow P., et al.: *A Toolkit for Redesign in Health Care.* Agency for Healthcare Research and Quality (AHRQ). Prepared by Denver Health under Contract No. 290-00-0014-7. AHRQ publication no. 05-0108-EF, Sep. 2005. http://www.ahrq.gov/qual/toolkit/ (accessed May 6, 2008).

6. Zidel T.G.: *A Lean Guide to Transforming Healthcare: How to Implement Lean Principles in Hospitals, Medical Offices, Clinics and Other Healthcare Organizations.* Milwaukee: ASQ Quality Press, 2006.

7. Martin K.: Lean thinking. Slide presentation given at Joint Commission Resources Ambulatory Care Conference, Chicago, Oct. 1, 2007.

8. Zidel T.G.: A Lean toolbox: Using Lean principles and techniques in healthcare. *J Healthc Qual* 28(1):W1-7–W1-15, 2006.

9. Agency for Healthcare Research and Quality: Creation and coordination of operational and evaluation structure. In *Managing and Evaluating Rapid-Cycle Process Improvements as Vehicles for Hospital System Redesign.* AHRQ Publication No. 07-0074-EF, Sep. 2007. http://www.ahrq.gov/qual/rapidcycle/ (accessed Mar. 17, 2008).

10. Printezis A., Gopalakrishnan M.: Current pulse: Can a production system reduce medical errors in health care? *Qual Manag Health Care* 16(3):226–238, 2007.

CHAPTER TWO

5S EVENT

5S Event At a Glance

TOOL DESCRIPTION	5S is a process of five steps to ensure that work areas are systematically kept clean and organized, waste is reduced, and visual communications are enhanced.
TOOL PURPOSE	• Provide a foundation for a Lean health care environment
	• Establish standardization of organization and orderliness
	• Provide a foundation for good work flow
	• Communicate a sense of pride and precision to customers, including staff and patients
WHO IS RESPONSIBLE?	A cross-functional Lean team usually initiates and monitors 5S implementation, but all employees will be responsible for participating in the process.
HOW LONG DOES IT TAKE?	The process takes 1–2 days for a full event or is broken into action items over a period of time.
STEPS IN THE PROCESS	1. Sort (*seiri*).
	2. Simplify or straighten (*seiton*).
	3. Sweep (*seiso*).
	4. Standardize (*seiketsu*).
	5. Sustain (*shitsuke*).

When everyone knows where everything is and how to access it, workplace organization and space management are more likely to keep staff and patients safe. At the start of a 5S event, each step of the process may take only a few hours, and then eventually it will take only a few minutes each day to keep going. The workers in a particular target area are crucial to making this project a success. This step-by-step tool provides a structured approach and an easy methodology for departmental (or individual) order (tidiness), organization, and cleanliness.

Key concepts for conducting 5S include advance planning to do the following[1]:

• Identify an authorizing manager
• Determine goals and targets
• Conduct a preworkshop audit to gather baseline data and design metrics for data collection
• Conduct actual walk-throughs of the unit or department
• Talk with people involved
• Observe processes as they stand
• Gather baseline data and design metrics for data collection
• Practice excellent communication

The 5S tool can deliver immediate payoffs in many areas. Among other benefits, 5S can do the following:

• Eliminate the necessity of hunting for information and supplies
• Reduce the probability of committing mistakes and errors

- Reduce the likelihood of some safety issues
- Increase productivity and maintain flow
- Improve quality
- Expedite response time
- Improve morale
- Alter the appearance of a unit or department to portray an organized and professional image
- Free up space and furniture (for example, file cabinets when less filing space is needed)

Steps in the Process

The five steps include creating a plan to accomplish the following tasks:

1. Sort the necessary from the unnecessary to free up space.
2. Simplify or straighten to create a standard layout.
3. Sweep (scrub/clean/shine/check) to make sure equipment is suitable for the job.
4. Standardize by conducting the first three S's at frequent intervals to maintain the workplace in a suitable condition.
5. Sustain the workplace by forming the habit of self-discipline in following the first four S's.

Step 1: Sort (*Seiri*)

This step sets the stage for all the other steps by getting rid of everything that is not being used or will not be used in the near future (within two weeks). Maintaining excess inventory and other means of hoarding is wasteful because it adds unnecessary inventory and supply costs and captures valuable space.

Everyone involved in the sorting process should agree on the mechanics of the process in advance of the tagging activity. For example, if several staff members are cleaning their cubicle spaces and filing areas, it will be beneficial to identify what each color tag means and where these tagged items will be centrally located during the sorting activity, to be reviewed and disposed of later. First, create an inspection sheet and tags to identify each item's name, date, and location. Tagged items can include equipment, supplies, instruments, tools, medications, other inventory, paperwork, files, furniture, even signs or posters on walls. All items tagged should be listed in a log so that they can be accounted for in financial documentation. In fact, the team should consider consulting financial staff within the organization before disposing of any items.

The first part of actual sorting is differentiating what is needed from what is not needed and moving tagged items to a predetermined, clear area. For needed items, determine whether items are needed frequently, occasionally, or only rarely. Then determine the disposition of each item, that is, whether it should be returned to its original place or stored elsewhere. For those items that are not needed, decide whether to dispose of each item (recycle or discard) and how to move it to its next destination. A plan of action should be put in place as well as a time line (by when) for moving items out of the work space. Other people may be recruited to help in this process, whether to help move large equipment out of the building or to another department or warehouse, or simply because others can be more objective about the usefulness of items and will not be likely to hang onto something for reasons of sentimentality or familiarity.

FIGURE 2-1. Red Tag

Department/ unit	**RED TAG**	Tag number

Category *(check one):* ❑ Equipment ❑ Office materials ❑ Medication ❑ Books
❑ Supplies ❑ Patient items ❑ Furniture ❑ Measuring instrument
❑ Other

Tag date:	Tagged by:

Classification *(check one):* ❑ Hazardous ❑ Non-hazardous

Item name:

Fixed asset code:	Serial #:

Quantity:	Value: $

Reason tagged: ❑ Not needed ❑ Beyond expiration date ❑ Borrowed ❑ Use unknown
(check one) ❑ Not used in 6 months ❑ Not used on unit ❑ Defective equipment

Disposition by – Authorized person's name:	Dept.:

Disposition by: ❑ Discard ❑ Move to storage ❑ Return to Lender ❑ Use
(check one) ❑ Repair ❑ Replace ❑ Move to holding area ❑ Other

The red tag strategy is used to sort items in the workplace.

Source: Zidel T.G.: *A Lean Guide to Transforming Healthcare: How to Implement Lean Principles in Hospitals, Medical Offices, Clinics and Other Healthcare Organizations.* Milwaukee: ASQ Quality Press, 2006, p. 76. Reprinted with permission from *A Lean Guide to Transforming Healthcare: How to Implement Lean Principles in Hospitals, Medical Offices, Clinics and Other Healthcare Organizations,* ASQ Quality Press © 2006 American Society for Quality. No further distribution allowed without permission.

Tasks in this step include the following:
1. Define the physical area that is being examined.
2. Create an inspection sheet and identify items that are not necessary.
3. Tag items that need to be shared with others or disposed of.
4. Move tagged items to a predetermined staging area.
5. Determine what is to be done with the items.

One method used to sort items in a workplace is described in the literature as the *red tag strategy.* An example red tag template is shown in Figure 2-1. Using this technique, everyone in the work area, and on all shifts, has an opportunity to review items of questionable need before another disposition may be made. The red tags may be professionally printed, computer generated, or made by hand on index cards or other heavy stock. The information on the tag includes the following:

- Date
- Name of the person tagging the item
- Item description
- Reason the item is being tagged
- Disposition
- Person who authorized the disposition

- Any code numbers
- Serial number

When reviewing tagged items, any employee can comment on the back of the tag when he or she has a question or concern. After all the employees in the unit or department have reviewed tagged items, authorized individuals submit items for disposition. All red tag items should be removed from the area prior to starting the next step.

Step 2: Simplify or Straighten (*Seiton*)

This step involves organizing all the needed items that remain after Step 1. The focus is on creating order and effective storage. Items should be arranged so that they can be easily located and easily returned to their storage places. The arrangement of items should be so intuitive and clear that even someone who does not work in

Tool Tangent

ORDER COMMUNICATES QUALITY AND SAFETY

At the Ambulatory Surgical Center at West Penn Hospital, part of the Pittsburgh Regional Hospital Initiative, staff used a 5S event to put a number of organization improvements into place[3]:

- Recliners were made available in the pre-evaluation area for patients who had difficulty climbing onto exam tables.
- Supply items in bedside cabinets were sorted to move seldom or never-used items to other hospital areas.
- Registration tasks (such as demographic verification or computer work) were done prior to patient arrival.
- Software errors in the blood bank's coding system were corrected so that paperwork was stopped from being sent to the wrong unit.

the area can locate the items. Items that are used often should be close at hand, and those used seldom can be farther away. This arrangement cuts down on wasteful hunting and searching motions. The more wasteful motion that is undertaken, the less time that can be allotted to value-added activities, foremost of which is anything affecting patient care. It has been estimated that nurses in particular spend only one third of their time taking care of patients and the other two thirds of their time "hunting, documenting, and clarifying."[2]

Label everything in the area. Use color coding liberally on labels, files, and signs. Everyone should be able to tell what is in a drawer, shelf, or cart at a quick glance. Some organizations use see-through doors and windows on cabinets. Make sure all staff can get to supplies and materials and that they are not piled up or obscured. Create a checklist for arranging similar items.

Create a standard layout that will allow staff to easily see if an item is out of place or not returned, and monitor the area to ensure that it stays standardized. For example, to help ensure that items are returned to their proper place, foam sheets can be cut to serve as a mold or template into which an item can be placed. Other aids to consider are colored duct tape, labels of all kinds, or any other means to make sure that things will stay in the "homes" where they are placed during the 5S process. Encourage staff to be creative here. Some sample straightening techniques are shown in Figure 2-2 on the next page.

Two particular areas that deserve extra attention to improve a work area's flow and structure are lighting and inventory issues.

Lighting Issues

One part of simplifying (straightening) is an assessment of the lighting and the availability of electrical outlets. Lighting should be neither too intense so that it enervates workers nor too dim so that items cannot be found or read easily. Lighting should be consistent for each work

---

FIGURE 2-2. Visual Organizational Controls

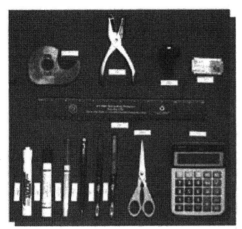

Example of a drawer organized with foam matting.

Example of colored tape used to ensure proper sequencing of manuals.

These two examples show how a visual cue such as a signal card, or kanban, *can help create and maintain order and thus cut down on time needed to locate needed supplies and resources. (a) Example of a drawer organized with foam matting; (b) example of colored tape used to ensure proper sequencing of manuals.*

Source: Zidel T.G.: *A Lean Guide to Transforming Healthcare: How to Implement Lean Principles in Hospitals, Medical Offices, Clinics and Other Healthcare Organizations.* Milwaukee: ASQ Quality Press, 2006, p. 78. Reprinted with permission from *A Lean Guide to Transforming Healthcare: How to Implement Lean Principles in Hospitals, Medical Offices, Clinics and Other Healthcare Organizations,* ASQ Quality Press © 2006 American Society for Quality. No further distribution allowed without permission.

---

space and each worker. Fuses, cables, and cords should be safe and, if possible, energy efficient. Sometimes, brighter lighting can make the difference to a decreased cycle time.

Inventory Issues

Poorly designed and executed inventory systems sometimes indicate that too much is being stored for too long, creating expired items that create costly waste. The opposite end of the inventory spectrum is when stock runs out, creating wasteful and sometimes dangerous circumstances. Both of these extremes should be eliminated. Following the Lean concept of just in time (described in the related "Lean Concept: Just-in-Time" sidebar), supplies should be available at the proper time and in the proper amount, and be aligned with a pull system of stakeholder demand.

Creating bin systems is one way to solve the problem of running out and facing an inventory crisis. When two or three bins are allotted to a particular item—such as syringes—then an empty bin signals it is time to order stock. Bar coding or other efficient communication systems may assist storerooms to connect with inventory. The related Lean tool of *kanban* (signal

Lean Concept
JUST-IN-TIME

Just-in-time (JIT) is the Lean concept of supplying staff or the customer with the precisely needed product or service, in the right amount, at the requested time, whenever it is needed, every time it is needed. This concept saves wastes of motion, cost, and quality and increases safety. JIT is synonymous with continuous flow and is one of the main features of a pull system.

Pull systems are driven by the needs of the downstream lines. By specifying value, identifying the value stream, and creating flow, Lean thinking allows pull to take place. The patients pull the product along rather than having the marketplace push it onto them on the organization's timetable. Pull offers more flexibility and accommodates changes in customer demand. In contrast, the work completed in push systems is driven by the output of the preceding lines. It is work that is pushed along regardless of need or request. It involves providing a service or product in anticipation of a need and is often associated with high inventory and the risk of errors or higher error rates. In Lean, push must be eliminated and replaced with a pull system to facilitate a JIT flow.

JIT can be applied to supplies or to patients. An example of JIT applying to the patient is when a patient is transported to ancillary departments, such as radiography, and must wait an hour in the hall on a gurney or in a wheelchair, exposed to passersby. To provide value to the patient, JIT would serve to notify the caregiver when the patient can be transported, but not so early as to disrupt the x-ray department's work flow or cause the patient's dissatisfaction.

Tool Connection
KANBAN

Kanban is the Japanese word meaning "signal card" or "signal cards." A *kanban* is most commonly a visual card or other indicator, attached to supplies or equipment, that serves as a means of communicating to an upstream process precisely what is required at the specified time.

Kanban is a method of just-in-time production that often uses standard containers or lot sizes with a single card attached to each. *Kanban* can be used for surgical kits, inventories, equipment, and forms. *Kanban* signals can include an empty bin, an alarm or alert, a light, or an outline painted on the floor.

A *kanban* should be inexpensive, convey its message clearly, and elicit a rapid response. A *kanban* might signal more than one department, the one who is the supplier and the one who will receive the delivery.

When *kanban* are used to point to needed supplies, a signal can be returned from the sender and receiver when the supplier has been notified, when the item has been sent, and when it has been delivered.

Kanban can be posted, painted, or labeled on any surface—on walls, ceiling, or floor—and in every size or shape, whatever works best. As with any sorting activity, integrate colors and designs to improve functioning.

Use *kanban* to prevent running out of items, depleting back stocks, and wasting time and movement. *Kanban* also can be especially valuable in improving patient flow.

card, as described in the "Tool Connection: Kanban" sidebar) is especially useful at this step. As shown in Figure 2-3 (next page), a *kanban* can alert staff that something is near depletion.

Step 3: Sweep (*Seiso*)

The third *S*, sweep, is also variously referred to as "shine" or "scrub." To maintain the area on a regular basis, conduct a "spring cleaning," and then create a cleaning plan for the area that could include daily and weekly tasks, such as cleaning the phone receiver or disposing of expired medications, or monthly tasks, such as monitoring for proper lighting. This step also calls for each individual to make a plan to clean his or her own area.

Some tips for executing this step follow:

- Some people find it helpful to take "before" photographs to help evaluate the risk and rough areas or issues.

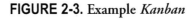

FIGURE 2-3. Example *Kanban*

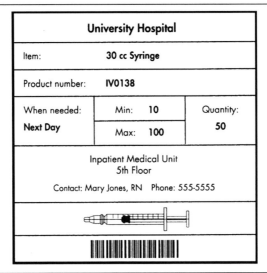

University Hospital		
Item:	30 cc Syringe	
Product number:	IV0138	
When needed: Next Day	Min: 10 Max: 100	Quantity: 50

Inpatient Medical Unit
5th Floor
Contact: Mary Jones, RN   Phone: 555-5555

This figure shows an example of a kanban *(signal card) used for replacing syringes in inventory.*

Source: Zidel T.G.: *A Lean Guide to Transforming Healthcare: How to Implement Lean Principles in Hospitals, Medical Offices, Clinics and Other Healthcare Organizations.* Milwaukee: ASQ Quality Press, 2006, p. 80. Reprinted with permission from *A Lean Guide to Transforming Healthcare: How to Implement Lean Principles in Hospitals, Medical Offices, Clinics and Other Healthcare Organizations,* ASQ Quality Press © 2006 American Society for Quality. No further distribution allowed without permission.

- Adopt some method to clean as a daily activity. Clean the work area or a section of the area before starting your job in the morning and at the close of day. Put aside a segment of time (for example, 15 minutes) for this activity each day.
- Inspect parts and places to ensure they are working.
- Find and design ways to prevent dirt accumulation and contamination.
- Clean equipment on a regular basis; perhaps share this task with coworkers.
- Identify and tag every item that causes contamination. Use the 5 Whys tool (see Chapter 4) or cause-and-effect methods to find the root causes of such contamination, and take appropriate corrective and preventive action.
- Keep checklists and a log of all places or areas to be improved and a "by when" date for when a solution will be identified.

Step 4: Standardize (*Seiketsu*)

Create guidelines to keep the area organized (the first *S*), orderly (the second *S*), and clean (the third *S*). Identify necessary tasks, who should perform them, how frequently the tasks should be performed, and supplies required. Create a five-minute checklist to help the assigned team members complete tasks on time. Keep standards in an easily visible place in the target area(s).

Backsliding into messes, disorganization, and unsightliness can happen easily unless preventive mechanisms are put in place. Dust and dirt accumulate each week, stains and spills happen and are left, and books and equipment remain where they are left when workers are busy.

Standardizing begins with ensuring that everyone in the work area knows what is expected of them. Simple plans for maintaining order and cleanliness should be posted and discussed with everyone. Plans may include removing items from top bins only or switching bins when the top one is empty. Create an agreement that the person who uses the last item in a bin initiates the call to replenish stock.

Keep these plans simple. Complicated procedures often are doomed to failure.

Everyone must be held accountable. Some people are natural housekeepers and some are not. But it is the rare individual who does not enjoy working in a clean and orderly environment. Some workers may help pull their colleagues along in the plan, inspiring them to follow the plan. For those who resist all such attempts and refuse to perform to these expectations, management might have to intervene to get them on board.

Tool Tangent
VISUAL INDICATOR FOR NUTRITION FOLLOW-UP

Although dietitians at one hospital were involved with designating the dietary regimens of patients with different medical conditions, such as those with heart disease, they had no clear means of follow-up either during hospitalizations or after discharge. Clarifying that the patient was the ultimate customer and that the dietitians' objective was not only to recommend the proper foods and beverages but also to teach patients how to select foods wisely, one remedy the dieticians instituted was the use of a visual indicator, "heart healthy" stickers, for instance, which helped standardize communication among patients, the dietary counselors, and physicians.

Step 5: Sustain (*Shitsuke*)

Of all the steps, the fifth and final step—sustain—is the most challenging. Maintaining self-discipline to sustain achievements over time will require the involvement of management, but in a supportive way. Communications should be positive and inspiring. Keep up training and discussions, issue stimulating memos and reminders, and post colorful, well-designed signs in conspicuous places. Get feedback continually from those involved to learn how to keep improving the process. Keep it fun, brainstorm themes, and be creative.

Try these strategies to design a learning environment that will support self-discipline:

- Create a training chart with specific topics directly related to staff members' jobs.
- Create checklists in different work areas to keep track of items and activities that must be maintained.
- Conduct regular audits of achievements to ensure that standards are maintained.
- Conduct regular education.
- Practice regular communication to keep up the success.
- Find ways to inspire staff to self-energize the process by means of visual displays, rewards, office/unit/ward competitions, showcases, and the like.
- Make recognition and acknowledgment part of the ongoing process.

To ensure that everyone is developing habits to maintain order and organization and that the *kanban* are working as they should, put multiple reminders in place during the first month to keep up momentum for maintaining self-discipline.

The goal of 5S is not only to clean and organize and to sustain that order, but moreover to eliminate waste and non-value-added work (as shown in Figure 2-4 on the next page).

FIGURE 2-4. Before and After

BEFORE

AFTER

Prior to a 5S workshop, supplies in this pathology department at Shadyside Hospital in Pittsburgh were stored on the floor and countertops, as shown in the "Before" photo. Afterward, countertops were clear and drawers and cabinets labeled with their contents, as shown in the two "After" photos.

Source: Lean Enterprise Institute. *The Anatomy of Innovation.* 2004. http://www.ihi.org/IHI/Topics/MedicalSurgicalCare/ MedicalSurgicalCareGeneral/Literature/TheAnatomyofInnovation.htm (accessed Mar. 21, 2008), reproduced by permission.

Relevant Health Care Applications

Health care organizations use 5S in numerous ways. One hospital operating room used 5S to redesign its supply chain to ensure a steady flow of supplies delivered directly from the distributor.[1] A general multispecialty inpatient surgery team used 5S to redesign processes for surgical case cart staging and equipment storage. *Kanban,* such as colored laminated cards, have been used to guide patients through the registration process, signaling when they have arrived for their appointment, are ready for the physician, or are in treatment. A pull just-in-time approach to inventory is used to cut the supply of expensive chemotherapy drugs so that they are available when they are needed as opposed to being stockpiled months in advance. A dry-erase board in each patient's room posts a checklist of medical and functional goals (such as self-bathing and feeding) for the patient and staff both to see when patients will be designated for discharge.

In one workplace redesign that included standardized patient room layout and equipment and stocking patient supplies at the point of use, outcomes included a 43% reduction in overall waste, a 30% increase in care-related activities, a 27% increase in bedside time, and a 12% decrease in steps defined as wasted motion.[4]

5S can be used in other less obvious areas, such as staff break rooms, locker rooms, housekeeping areas, closets, or waiting areas. Any storeroom or warehouse is also appropriate for use of the 5S tool.

REFERENCES

1. McAuliffe J.: Practicing "wasteology" in the OR. *OR Manager* 23(3):10–15, 2007.

2. Zidel T.G.: *A Lean Guide to Transforming Healthcare: How to Implement Lean Principles in Hospitals, Medical Offices, Clinics and Other Healthcare Organizations.* Milwaukee: ASQ Quality Press, 2006, p. 1.

3. Pittsburgh Regional Hospital Initiative puts new spin on improving healthcare quality. *Qual Lett Healthc Lead,* Nov. 2002.

4. Healthcare Performance Partners: *Nursing Team Redesigns Floor and Eliminates Waste.* http://leanhealthcareperformance.com/healthcaredocuments/HPPCaseStudyNursingRedesign.pdf (accessed Mar. 18, 2008).

CHAPTER THREE

KAIZEN EVENT

Kaizen At a Glance

TOOL DESCRIPTION	In Japanese, *kai* means "to take apart" and *zen* means "to make good." As a whole, the word *kaizen* may be translated as "improvement." A *kaizen* event is an intensive workshop to tackle a focused problem and to set a path for continuous incremental improvement over a sustained period.
TOOL PURPOSE	• Ensure long-term success through continuous, evolutionary change. • Quickly implement Lean tools to eliminate waste and non-value-added work.
WHO IS RESPONSIBLE?	Cross-functional team, generally no more than 10 people
HOW LONG DOES IT TAKE?	1–2 months from planning to follow-up 3–5 days to accomplish the on-site *kaizen* workshop
STEPS IN THE PROCESS	1. Planning phase (2–4 weeks) 2. *Kaizen* workshop (2–5 days) 3. Follow-up phase (3–4 weeks)

Lean experts believe that the only sustainable process is one in which the participants are fully enrolled—that is, the participants need to view the process as a whole and understand its logic. One way to accomplish such engagement is to hold a *kaizen* event. *Kaizen* events employ aggressive, measurable objectives to implement continuous, incremental improvement with the goal of increasing value and reducing waste.

The most effective *kaizen* event includes the planning phase and follow-up phases as well as the workshop phase. Different perspectives see different problems and different solutions. Include people with many roles and responsibilities in the team.

One hundred percent focus is imperative for a successful *kaizen* event. Breaking the flow of concentration requires start-up time to get back up to speed. Thus, in a *kaizen* event (often called a rapid process improvement, or RPI, event), a cross-functional team is sequestered for the duration of the event, generally scheduled for two to five days. The team first defines and exploits the technical boundaries of existing resources to engage creativity before moving to "capital thinking,"[1] that is, solving problems by implementing change using supplies or services that add on new costs. Rapid change can be implemented during the event, which spurs on greater momentum.

Steps in the Process

The *kaizen* event includes a planning phase, the *kaizen* workshop itself, and a follow-up phase.

Planning Phase

Current state and future state value stream maps (described in Chapter 1) can identify problems or areas in which waste can be eliminated. This information can be used as a road map for the *kaizen* event. The current state map is metric based and makes use of spaghetti diagrams to plot location of equipment, motion, and the flow of activity.

Assemble the multidisciplinary *kaizen* team and identify a team project champion from leadership. An event can be successful only if management, through the project champion, is fully supportive and staff members fully cooperate. Champions orchestrate the most effective deployment of the Lean program by providing the support the team will need to execute a successful implementation plan. The champion assumes the role of adviser to leadership teams on the most appropriate rollout plan in their area and stretches the organization toward rapid improvement implementation.

Kaizen (Rapid Process Improvement) Workshop

The team composition for the workshop will be determined by the scope of the event. A cross-functional team should include no more than 10 members for maximum effectiveness. The assembled team could include a combination of process workers, upstream suppliers (internal), downstream customers (internal), subject matter

Tool Connection
SPAGHETTI DIAGRAM

A spaghetti diagram helps highlight the movement of people, makes waste of motion and transportation obvious, and allows the viewer to observe where the operator moves and why. This tool helps establish the optimum layout for a department or unit based on the distances traveled by patients, staff, or products (for example, x-rays). The complexity and routes of movement facilitate reducing motion in a commonsense fashion, which creates order, saves time, and increases value. String may be used to make a spaghetti diagram, pinning the beginning of the string at the starting point on the board or display and routing it along a series of connections. (For an example of a spaghetti diagram, see Figure 9-2 in Chapter 9.)

Lean Concept
UPSTREAM AND DOWNSTREAM

The term *upstream* designates a provider of a service or work unit required from a customer. *Downstream* designates a customer, patient, provider, and/or process that requires an upstream service and is paying for it.

experts, someone to apply "outside eyes," external suppliers or contractors, external customers, and if relevant or needed, administrative support, union representatives, maintenance representatives, and management.

Create a project checklist or milestone worksheet to detail improvement activities. The scope of the event defines what the team will focus on. It may also help to define what will not be undertaken. It is important to be realistic about what can be accomplished. Plans that are too aggressive may create rushing and frustration. Changes may have to be introduced gradually through pilot projects. Define the scope of the event in as much detail as possible without suppressing the capacity of team leaders to be flexible during the event. Remember that the customer is always the focus of the work.

Careful planning means that no time will be wasted and event objectives will be met. Consider all needs thoroughly, including communications, participants (including accommodations for those whose daily responsibilities must be covered), requirements (including special equipment), food, setting (such as hotel or conference rooms), and paperwork. Before the event, post a *kaizen* event charter, which may include, but not be limited to, current-state issues, team members, event goals and objectives, potential deliverables, possible obstacles, on-call support, and needed approvals. To develop the charter, solicit ideas from stakeholders, including upstream and downstream workers.

Objectives, which should be considered carefully and be agreed on by the leadership team, should be, once again, SMART: *s*pecific, *m*easurable, *a*ttainable, *r*elevant, and *t*ime limited. Choose issues that relay urgency and inspiration. Conservative objectives will result in limited accomplishments with only reasonable outcomes. In total, the planning phase takes four to six weeks prior to the *kaizen* workshop.

In general, follow these steps for conducting the workshop:

1. Train the team in Lean. Training may be incorporated into the workshop at various times. There may be a couple of hours of training on the first day and little bits or "chunks" on some of each of the first few subsequent days.
2. Begin by applying 5S and using the related tools to observe the area of focus and to gather data on cycle times. (See Chapter 2 for discussion of 5S.)
3. Form subgroups to brainstorm improvements for the event.
4. Implement improvements, create specific tasks, and identify follow-up tasks.

Generally, during a *kaizen* event, the current state map and the root causes of problems and barriers to flow are analyzed during the 5S process. Depending on the area of focus, some groups complete the current state map before the workshop, and some choose to accomplish walk-throughs before the workshop and then complete analyses and the future state map at the *kaizen* workshop. *Kaizen* events are about "try-storming," that is, trying new ideas quickly, not just stopping at brainstorming. *Kaizen* events are focused on creative solutions without additional costs.

A few other tools can provide data that are useful in a *kaizen* workshop; one of them is the Pareto chart, presented in Figure 3-1 on the next page. Numerous sheets and forms can help keep organization on track during the event. When the *kaizen* workshop is held, an execution checklist, such as the example in Figure 3-2 on page 50, can help keep steps and processes on track.

PARETO CHART

Also called the *Pareto diagram* or *Pareto analysis*, a *Pareto chart* is a bar graph that displays the activities being studied in order from largest to smallest. This tool is helpful to represent the activities in which staff members are involved and in depicting the ranking of those activities. The lengths of the bars represent frequency or cost (time or money) and are arranged with the longest bars on the left and the shortest on the right. The Pareto chart is useful when analyzing the frequency of problems or causes in a process, when wanting to focus on the most significant problems or concerns, when analyzing broad causes by looking at their specific components, and when communicating about data.

For the *kaizen* workshop to move smoothly, a number of rules should be followed. Table 3-1 on page 51 provides an example of the set of commandments that could be provided to a *kaizen* team to produce a smooth and successful experience. Assigning leadership and facilitation roles to team members is also crucial, as discussed in the following sections.

Facilitator

The event facilitator, or *sensei,* must be someone who is highly trained in Lean tools, root cause analysis, project and time management, team building and facilitation, and human relationships. The facilitator must be a natural magnet for respect by how he or she handles him- or herself and the subject. The *sensei* should feel

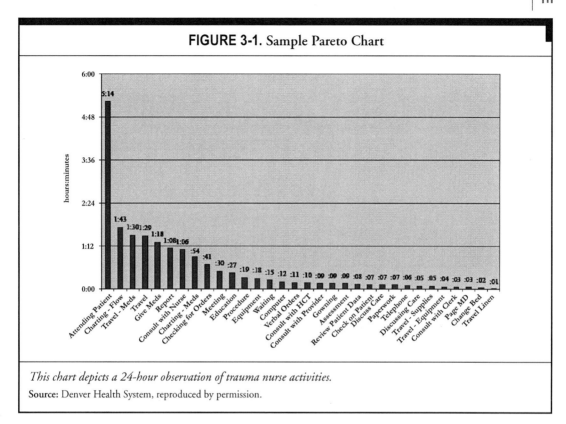

FIGURE 3-1. Sample Pareto Chart

This chart depicts a 24-hour observation of trauma nurse activities.

Source: Denver Health System, reproduced by permission.

comfortable removing obstacles in the way of outcomes and should be an authority who can reach out to both leadership and team members. The best facilitators have an upbeat, positive outlook; they are passionate about Lean and its potential and are masters of conflict resolution.

Team Leaders

An effective team leader is key to achievements during the event and afterward. An effective leader is organized, has good leadership skills including communication, is committed to and well versed in the Lean philosophy, and has a track record for getting things done efficiently.

The involvement of other upper management during the *kaizen* event is also extremely important. Some leaders will be integrally involved in the event, but others may be "on call" or might drop in to survey progress every day or two. The stance that leadership takes during the process is crucial.

As the team presents its ideas, leaders can query the team members about areas they may not have thought of or implications of their solutions. The leaders' role is not to say "no" but to ask, "Have you considered this?" or "How would we handle this?" or "What if ...?" The team members strengthen their resolve to move forward as they learn to identify and eliminate waste and as they continually remind themselves that their goal is to improve value for the patient (the customer). Leaders who maintain their focus on the strategic plan and performance monitoring add value to the process. Their role is to help keep the team "out of the weeds."[1]

FIGURE 3-2. *Kaizen* Event Execution Checklist

Kaizen Event
Execution Checklist

Spell-check Sheet			
Value Stream Champion			**Event Name**
Facilitator			**Event Dates**

	☑	N/A*	Description
1	☐	☐	Conduct Event kick-off / review Event Charter with team.
2	☐	☐	Conduct ice breaker / team introduction(s).
3	☐	☐	Deliver Lean and Kaizen Event overview training (if not delivered prior to event); review Event Charter.
4	☐	☐	Document current state (before) conditions (MBPM, relevant metrics, spaghetti diagram, photos, etc.).
5	☐	☐	Summarize / analyze current state metrics.
6	☐	☐	Brainstorm session re: possible improvements.
7	☐	☐	Prioritize improvements.
8	☐	☐	Create standard work, check lists, visual aids, etc. to document the new process.
9	☐	☐	Prepare for interim briefings.
10	☐	☐	Prepare agenda for following day (end of each day).
11	☐	☐	Identify who will need to be trained on new process and schedule training (held *during* the event).
12	☐	☐	Create and document future state / "after" conditions and calculate projected percent improvement.
13	☐	☐	Deliver training to those who perform or are affected by improved process.
14	☐	☐	Create 30-Day List, assign owners and deadlines for completion.
15	☐	☐	Create Sustainability Plan.
16	☐	☐	Complete Kaizen Event Report.
17	☐	☐	Assign ownership for tasks on "Post-Event Activities" list.
18	☐	☐	Schedule weekly follow-up meetings (for first four weeks following the Event).
19	☐	☐	Schedule 30-day audit.
20	☐	☐	Take team photos.
21	☐	☐	Prepare and deliver final presentation.
22	☐	☐	Recognize participants via certificates, shirts, gift cards, handshake from leadership, etc.
23	☐	☐	
24	☐	☐	
25	☐	☐	

This execution checklist can help keep steps and processes on track.

Source: Martin K., Osterling M.: *Kaizen Event Planner: Achieving Rapid Improvement in Office, Service, and Technical Environments.* New York: Productivity Press, 2007, p. 106, reproduced by permission.

Follow-up Phase

In general, the follow-up phase of the workshop consists of the following steps:

1. Report results to the project champion.
2. Continue to implement ideas from the event.
3. Create standards for the improved processes.
4. Submit regular status reports to the champion.
5. Create a sustainability plan that includes who controls the project at each project step, who will make what changes, and who will communicate to others that become involved.
6. Submit a final report when the *kaizen* event is complete.

Dropping the planning and follow-up phases in favor of holding a single *kaizen* workshop is a common temptation. Outlining lists and worksheets and scheduling follow-up meetings and audits ahead of time will reduce the chance of that happening. Useful tools include a *kaizen* event sustainability plan (Figure 3-3, page 52), a postevent activities list (Figure 3-4, page 53), and a *kaizen* event report (Figure 3-5, page 54).

TABLE 3-1. *Kaizen* Commandments

Behavior Oriented

1. The team starts and ends the day together.
2. Being on time is critical.
3. Cell phones, BlackBerries®, and other communication devices must be turned off or placed on 100% silent (no vibration) mode.
4. No interruptions.
5. Team stays in the room.
6. Avoid scope creep; keep focused on event objectives and work within predetermined event boundaries.

Communication Oriented

7. Finger pointing has no place. *Kaizen* Central is a blame-free zone.
8. No veto power from outside the team.
9. No silent objectors.
10. One conversation at a time.
11. What's said in the room, stays in the room.
12. It's okay (and encouraged) to disagree; it's not okay to be disagreeable.

Mindset/Philosophy Oriented

13. Rank has no privilege.
14. Think "creativity before capital."
15. Ask "Why?" and "What if?" and "How could we?"
16. Think "yes, if ..." instead of "no, because ..."
17. Eliminate "can't" from your vocabulary.
18. Seek the wisdom of ten rather than the knowledge of one.
19. All ideas are worthy of consideration.
20. Keep an open mind.
21. Improvements implemented today are better than planning to implement in the future.
22. Abandon departmental/functional/siloed thinking.
23. Stay focused on customer-defined value.
24. Focus on how the results are achieved, not just the results.

Source: Martin K., Osterling M.: *Kaizen Event Planner: Achieving Rapid Improvement in Office, Service, and Technical Environments.* New York: Productivity Press, 2007, p. 111, reproduced by permission.

FIGURE 3-3. *Kaizen* Event Sustainability Plan

Kaizen Event Sustainability Plan

Executive Sponsor		Event Name	
Value Stream Champion		Event Dates	
Facilitator		"Go Live" Date	
Team Lead		"Go Live" Location	
30-Day Audit Date		60-Day Audit Date	

Communication / Training

Row	Requirement	Plan
10	Provide training for those who missed initial training.	*Who will deliver it and when?*
11	Integrate new process into ongoing department training.	*Who leads identification of training that need to be updated (ongoing and for new employees), when will training be in place?*
12	Update Value Stream Map.	*Which value streammap(s) need to updated, who will do it and when?*
13	Update training records to reflect who has been trained.	*Who maintains training records?*
14	Communicate to affected parties who were not advised during event.	*Who communicates? How? To whom?*
15	Post the Event Report, 30-Day List, Sustainability Plan.	*Who is accountable? Where posted?*
16	Update SOPs and other ISO or regulatory documents impacted by changes.	*Who identifies relevant documentation? Who updates it? By when?*
17	Communicate and post 30-day and 60-day audit results.	*Who communicates? How? To whom? Where posted?*
18	Communicate audit results to stakeholders and leadership team.	*Who communicates? Via what medium?*
19	Communicate process performance measurement plan.	*How is process performance going to be communicated to workers? Are additional visuals needed? Who owns this activity?*
20		
21		
22		
23		

Monitoring / Measurement

Row	Requirement	Plan
26	Identify process owner.	*Who has ultimate responsibility for how process is perfoming?*
27	Observe process **one day** after event is over. Assure that workers are performing the process correctly and see if there are problems.	*Which Kaizen teammember(s) will observe the process? How will they communicate results? Who is responsible for adjusting the process, if needed?*
28	Put in place key metrics to measure process performance; post performance.	*Who is responsible for finalizing process perfomance metrics and by when?*
29	Observe process **one week** after event is over. Assure that workers are performing the process correctly and see if there are problems.	*Which teammember(s) will observe the process? When? How will they communicate results? Who is responsible for adjusting the process, if needed?*
30	Define corrective actions required if new process not being followed (rewards and consequences).	*Will there be rewards / recognition for sustaining improvements? What will corrective actions be if workers are not following new process? Who will enforce these rewards/consequences?*
31	Monitor process performance frequently; post results put continuous improvement plans in place.	*Who monitors process perfomance on an ongoing basis? Who communicates the results? When, to whom in what format?*
32	Conduct 30-day audit.	*Who will lead the audit & when? How will results be communicated? To whom? How will the process be adjusted, if needed? What's the plan for continuous improvement?*
33	Conduct 60-day audit.	*Who will lead the audit & when? How will results be communicated? To whom? How will the process be adjusted, if needed? What's the plan for continuous improvement?*
34		
35		
36		
37		

Value Stream Champion		Process Owner	
Name		Name	
Signature		Signature	
Date		Date	

The sustainability plan sheet, the first page of which is shown here, helps ensure that the workshop end does not signal activity's end.

Source: Martin K., Osterling M.: *Kaizen Event Planner: Achieving Rapid Improvement in Office, Service, and Technical Environments.* New York: Productivity Press, 2007, p. 173, reproduced by permission.

FIGURE 3-4. Postevent Activities Checklist

	☑	Task	Due Date	Owner	Comments
10	☐	Share "later" ideas on Ideas List with relevant leadership and continuous improvement staff.			
11	☐	Modify process further if it needs additional improvement.			
12	☐				
13	☐				
2 Weeks After					
14	☐	Meet with team to assess progress on 30-Day List.			
15	☐	Measure process to validate projected future state metrics.			
16	☐	Adjust process, if needed.			
17	☐	Interview process workers and those affected by the improvements to assess success and need for future improvements.			
18	☐				
19	☐				
3 Weeks After					
20	☐	Meet with team to assess progress on 30-Day List.			
21	☐	Measure process to validate projected future state metrics.			
22	☐	Adjust process, if needed.			
23	☐	Determine data needs required for further current state analysis and future state design. Assign accountability for measurement and timeframes for delivery.			
24	☐	Prepare for 30-day audit.			
25	☐				
26	☐				
4 Weeks After					
27	☐	Meet with team to assess progress on 30-Day List.			
28	☐	Conduct 30-day audit.			
29	☐	Adjust process and/or take corrective action, if needed.			
30	☐	Conduct 30-day audit briefing and post audit results.			
31	☐	Schedule 60-day audit.			
32	☐				
33	☐				

Kaizen Event Post-Event Activities — Executive Sponsor, Value Stream Champion, Facilitator, Coordinator / Event Name, Event Dates

As the partial view of this checklist shows, you can continue to monitor activities following the kaizen *event at one week after, two weeks after, and so on.*

Source: Martin K., Osterling M.: *Kaizen Event Planner: Achieving Rapid Improvement in Office, Service, and Technical Environments.* New York: Productivity Press, 2007, p. 187, reproduced by permission.

Relevant Health Care Applications

Kaizen has been used to achieve a number of objectives, including reducing wait times and streamlining patient flow. One hospital initiated a new system for bedside registration to allow quicker treatment. *Kaizen* has been used to solve the problem of eliminating lab specimen relabeling, which can improve productivity and reduce potentials for error. The burden of paperwork has been reduced in a daily operating room (OR) schedule and in surgical instrument use and processing. A transition unit has helped discharge flow issues, which frees up beds throughout the hospital.[2]

FIGURE 3-5. *Kaizen* Event Report

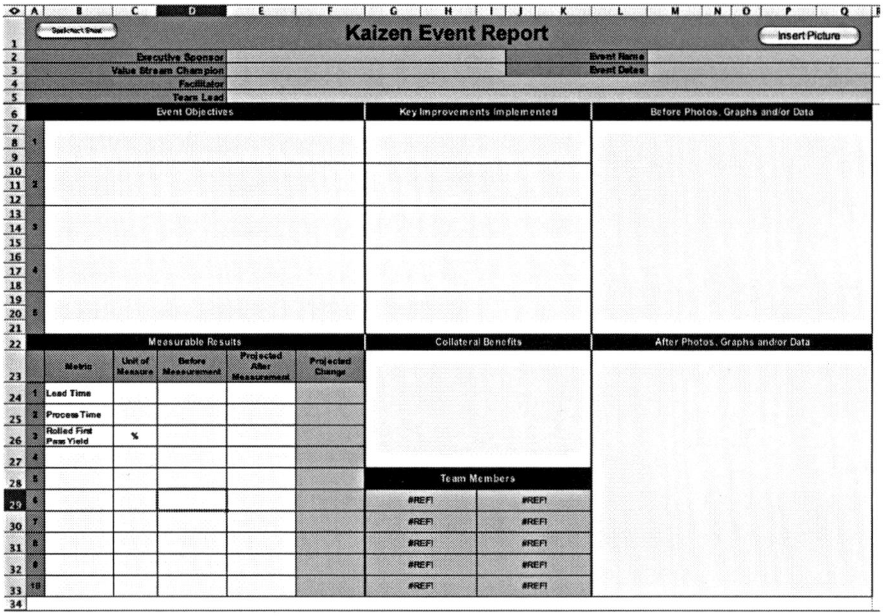

This example report template helps summarize the kaizen *event's actions and projected results.*

Source: Martin K., Osterling M.: *Kaizen Event Planner: Achieving Rapid Improvement in Office, Service, and Technical Environments.* New York: Productivity Press, 2007, p. 170, reproduced by permission.

Kaizen presents an effective tool for improving clinical processes. In one example, a team of 15 people at the University of Iowa Hospitals and Clinics in Iowa City used Lean Six Sigma *kaizen* over five days to eliminate non-value-added activities in radiology computed tomography (CT) scanning processes.[3] The goal was to eliminate a two-week lag time in scheduling outpatients for CT scans, which sometimes made it necessary to refer patients to other facilities. The measurable objectives to achieve that goal included increasing throughput of the CT scanners by 5% and reducing overall patient cycle time by 15%. At the end of the project, the skills designated as value-added were those requiring a technologist and the steps contributing directly to the operation of the CT unit. The recommended process changes resulted in the following:

- Sixty fewer calls to the command center (30% reduction)
- A 31% increase in patient throughput, a 30% reduction in total *takt* time (from 31 minutes to 21 minutes)
- A 33% reduction in patient experience time (from 114 minutes to 76 minutes)
- A 91% reduction in travel time for technicians and a 50% reduction in travel time for prep personnel

The results showed that more than 3,000 additional CT cases could be handled each year, which translated to an annual revenue increase of $750,000.

The University of Washington Medical Center in Seattle[4] has used RPI and Lean thinking in its 14-room main OR and 11-room outpatient surgical center. Some of the center's innovations included a "block doc" (an attending anesthesiologist teamed with a resident and registered nurse to start nerve blocks on patients in the preoperative holding area), an OR script that maps each team member's activities for cases, a green card *kanban* signaling when a patient in the holding area is ready for the OR, and a time-saving anesthesia drug distribution process. Using this method, a 1.4-day savings in length of stay was subsequently demonstrated for orthopedic patients who had regional nerve blocks.

The staff at Virginia Mason Medical Center in Seattle[5] used *kaizen* to improve nonclinical processes in health care organizations. Staff members have found ways to reduce the distances they walk by 34 miles and the distances supplies must travel to reach destinations by 70 miles. Inventory costs have been slashed by 51% and lead times reduced by 708 days. Productivity gains have freed up 77 full-time equivalents (FTEs)—a 44% reduction. Defects have declined by 47%. One-week *kaizen* events have resulted in a cumulative savings of at least $12 million budgeted for health system improvements such as a cancer center.

After a *kaizen* event for labor and delivery staff in one 150-bed hospital unit, the team did the following[6]:

- Created standardized work and a standard operating procedure to incorporate medication-eligibility notification into the admissions process
- Created a new fax-back system and reestablished an existing pager protocol to stabilize communication between nursing stations and recovery room environmental teams
- Created standardized work practices requiring advance scheduling of discharge procedures and 24-hour advance initiation of those procedures and related paperwork
- Acquired rolling admissions computers and reconfigured a model labor-and-delivery room for efficiency and bedside care
- Realized $94,000 in annual savings from process improvements and other additional long-term savings
- Reduced patient wait times, including a one-hour reduction for admissions, a one-hour reduction for medication, and a three-hour reduction for discharge
- Reduced nurses' per-patient travel to less than a mile, resulting in 33% more bedside time per nurse and reconfiguration of all labor-and-delivery rooms for better mother/baby care

In the health care setting, *kaizen* events have been used for a wide range of operations and support processes as well.[7] Events lasting from two days to several weeks have been used in the areas of admission and discharge; triage; catheterization and other laboratories, including blood banks; and outpatient clinics. In the area of finance, *kaizen* events have helped staff make improvements in bad debt, emergency department collections, and inpatient coding. *Kaizen* events have helped streamline and improve processes in surgical areas and have helped improve the risk-laden area of patient transfer (handoffs). For example, successful hospital projects have included moving all OR instruments from OR to sterile processing within one week. In another project, lab specimen turnover time was reduced from over 2 hours to 42 minutes.

Another *kaizen* effort at the University of Washington in Seattle was conducted to explore how to free up nonoperative time (the time from closing the patient's incision to beginning incision on the next scheduled patient) in thoracic surgery. The objectives were to add more cases, increase billable hours, and reduce stress on surgical staff. Nonoperative time was reduced by 13% (an average of 14 minutes per case), turnover process time between patients was reduced by 560%; and the OR team's average travel distance was reduced by 46% (from 2.4 miles to 1.3 miles).

REFERENCES

1. Martin K.: Lean thinking. Slide presentation given at Joint Commission Resources Ambulatory Care Conference, Chicago, Oct. 1, 2007.

2. Kolodziej J.H.: The Lean team. *Mich Health Hosp* 37:24–26, Jan.–Feb. 2001.

3. Bahensky J.A., Roe J., Bolton R.: Lean Sigma: Will it work for healthcare? *J Healthc Inf Manag* 19(1):39–44, 2005.

4. McAuliffe J.: Practicing "wasteology" in the OR. *OR Manager* 23(3):10–15, 2007.

5. Weber D.O.: Toyota-style management drives Virginia Mason. *Physician Exec* 32(1):12–17, 2006.

6. Healthcare Performance Partners: *Labor and Delivery Realize Annualized Savings of $94K & Increases Bedside Time by 33%.* http://leanhealthcareperformance.com/healthcaredocuments/HPPCaseStudyL&D.pdf (accessed Oct. 21, 2008).

7. Advisory Board Company: *"Lean" Models at Health Care Institutions.* Original inquiry brief. Washington, DC: Marketing and Planning Leadership Council, Dec. 19, 2006.

CHAPTER FOUR

ERROR PROOFING

Error Proofing At a Glance

TOOL DESCRIPTION	Error proofing is a method of identifying processes that are most likely to generate mistakes and defects and revising them to reduce risk.
TOOL PURPOSE	Prevent and eliminate mistakes to achieve zero defects
WHO IS RESPONSIBLE?	Cross-functional Lean team
HOW LONG DOES IT TAKE?	Data collection and analysis may take a week or more.
	Verifying effectiveness usually takes a few hours to days.
STEPS IN THE PROCESS	1. Perceive.
	2. Analyze.
	3. Standardize.
	4. Alert.
	5. Error proof.

Steven Spear, senior fellow at the Institute for Healthcare Improvement, puts it very well when he describes the problem of medical errors and mistakes: Out of approximately 33.6 million hospitalizations in the United States each year, as many as 88 people out of every 1,000 will suffer injury or illness as a consequence of treatment, and about 6 of them will die. Every 15 to 20 minutes, 5 to 7 patients will die as a result of medical errors and infections acquired in health care organizations, and 85 to 133 patients will be injured.[1]

It has also been estimated that for every death due to a medical error, there were 10 injuries and 100 instances where injury was prevented.[2] Most accidents are avoided, but these "near misses" can provide rich information to lead to systems improvements. The use of Lean thinking can rout out those pitfalls to prevent harm and sustain patient safety and quality.

Error proofing, known in Japanese as *poka-yoke,* applies concepts that can prevent mistakes in a process. Error proofing specifically targets processes that are most likely to generate mistakes. It seeks out risky conditions and revises them to reduce risk. It encompasses the ideas from knowledgeable staff to improve a process's or area's flow.

Clinical examples of error-prone processes include medication administration errors, wrong-site surgery, incorrect patient identification, delays to diagnosis, equipment and device visual controls, and hand washing/infection control. As Figure 4-1 on page 60 shows, patient safety alerts exemplify error proofing at its best, used in a clinical context. Efficiency and error prevention can be increased in the nonclinical areas of billing, medical records, and appointment scheduling. Error proofing reduces costs to the organization by reducing duplication of work and has advantages in saving waste in all areas.

Poka-yoke, or error proofing, is one of the main components of the Toyota Production System pioneer Shigeo Shingo's Zero Quality Control system. Anyone from a front-line nurse to a CEO can develop a *poka-yoke,* and it may look very different in various health care situations. Although it might seem obvious that anyone would want to stop and prevent mistakes from happening, skeptics might want to adhere to the familiar and what they have always done. Clear and concise data regarding variations and successful outcomes can help convert the reticent. Physicians are traditionally slow to change and may be especially reluctant to apply standard work and error proofing to the area of diagnostics. Clinicians who are approached to participate in Lean thinking may feel threatened and misinterpret Lean to mean that they are being asked to abandon the art of medicine, a cherished tenet. Quite the contrary: Methods, techniques, or equipment that can help standardize a high caliber of cognitive thinking and

Lean Concept
POKA-YOKE

A surgeon at a Rhode Island hospital operated on the wrong side of a man's brain after a computer tomography (CT) scan was inverted on an x-ray viewing box.[3] The patient's internal bleeding was on the right side of his brain, but the reversed scan made it look as if the bleeding was on the left. To compound the problem, the patient's incision site had not been marked with a pen as patient safety experts recommend. Error proofing at numerous points in the process could have prevented this problem.

promote greater clinical decision making will help increase diagnostic accuracy, reduce diagnostic delay, and improve treatment outcomes. Because Lean thinking is a fluid and additive concept best learned through experience, physicians may best see the benefits of Lean in all areas by being shown data from successful project outcomes.

Steps in the Process

Developing a Lean mistake-proofing system or device involves five steps. These steps are as follows:
1. Define the mistake or defect and identify the conditions that may lead to mistakes (perceive).
2. Conduct a root cause analysis (analyze).
3. Generate ideas to eliminate mistakes from happening in the future and to choose standards to follow (standardize).
4. Communicate the conditions that may cause errors (alert).
5. Develop and implement the methods or devices to prevent mistakes in the future (error proof).

These steps are described in the following sections.

Perceive

This stage involves changing your perspective to begin looking for errors and their sources. Then, adopt a habit of looking for opportunities to eliminate the sources. A mistake occurs when there is an intended planned action but the result fails to meet that intention. Also, the plan itself might be a source of the mistake. For example, not checking the identity of a patient before administering a drug is a faulty plan even though the intended action is to give the correct drug. Another example is using a weight on a patient's chart to calculate dosage, not realizing that the weight was estimated in the emergency department before the patient got to the floor.

Mistakes can result from situations that include confusing instructions or directions, errors in calculations or the data used for calculating, interruptions and distractions, carelessness, hurrying, poor training, and poor follow-up or confirmation techniques. Defects are the end result of mistakes from any source.

FIGURE 4-1. Patient Safety Alert Categories

Patient Safety Alert Category Examples*	
Airway Management ■ Self-extubation ■ Nebulizers not available during emergent treatment	**ID/Documentation/Consent** ■ Patient misidentification ■ Inadvertent release of patient health information ■ Missing consents
Blood/Blood Product ■ Delayed processing of blood orders	**Infection Control** ■ Sharps use ■ Inadequate setup of isolation precautions ■ Personal protective equipment
Care/Service Coordination ■ Communication and handoff issues among healthcare team	**Lab Specimen/Test** ■ Mislabeled specimens or unlabeled specimens
Diagnosis/Treatment ■ Delayed diagnosis or treatment resulting in worsened condition ■ Incorrect diagnosis or incorrect treatment provided to patient	**Line/Tube** ■ Infiltration of IV line ■ Incident related to order, preparation, insertion, or use of line or tube
Diagnostic Test ■ Missed orders or delayed communication of critical results	**Medication/IV Safety** ■ Delay in obtaining and administering STAT medications ■ Wrong medications administered to patient ■ Use of dangerous abbreviations on medication orders ■ Near miss of wrong dose or wrong medication
Environment ■ Staffing concerns ■ Malfunctioning equipment	**Safety/Security/Conduct** ■ Wandering patient ■ Allegations of sexual misconduct or assault ■ Verbal or physical abuse by staff ■ Drug-seeking behavior by staff
Employee ■ Injury from assisting patient ■ Noncompliance with fitness-for-duty requirements ■ Needlestick injury	**Skin/Tissue** ■ Trauma of skin/tissue, such as phlebitis, rashes
Fall ■ Patient fall in room, with or without assistance	**Surgery/Procedure** ■ Ordering, preparation, or performance of surgical procedure or anesthesia ■ Equipment not available for scheduled surgery

* ID, identification; IV, intravenous.

This chart shows examples of patient safety alert categories to assist in error proofing.

Source: Furman C., Caplan R.: Applying the Toyota Production System: Using a patient safety alert system to reduce error. *Jt Comm J Qual Patient Saf* 33:376–386, Jul. 2007.

The first step for the error-proofing team will be to define mistakes and to differentiate between mistakes and defects. Allow the group to brainstorm all the possible sources of mistakes. Next, give examples of these mistakes and defects from the work environment. Then, categorize the different types of mistakes (rushing, poor directions, etc.).

Analyze

The team will then work together to identify the defect, the resulting condition, and where the defect is found. You must be able to identify and describe the problem or defect in depth, including the rate of defect over time. Documenting the details of the mistake may provide a means for a group to discuss the mistake or defect. Ask the following questions as you record the error:

Lean Concept
ANALYSIS OF A DEFECT

Ventilator associated pneumonia (VAP) is a source of morbidity, complications, and cost at hospitals across the United States. Frequent hand washing by providers and keeping the patient's head elevated can help reduce the incidence of this complication, but hand washing often is not performed regularly enough. A registered nurse who was a member of the Mississippi delegation to the Patient Safety Improvement Corps believed it was important to be able to determine whether a patient's bed was raised to the necessary 30-degree angle to help prevent VAP. She perceived that a large, bright visual control could help nurses easily see from outside a room in the intensive care unit (ICU) when the bed was at the correct angle and would save their having to continually enter the room to check the ventilator gauge. If the sign is ever flat or straight up, they can see from a distance that the bed is not in the optimal position. The ability to detect those variations has been assimilated into the ICU nursing culture.[4]

FIGURE 4-2. *Kanban* Showing Bed Position

A bright-colored sign (kanban) *provides a visual control to determine the correct bed position.*

Source: Grout J.: *Mistake-Proofing the Design of Health Care Processes.* AHRQ Publication No. 07-0020. Agency for Healthcare Research and Quality (AHRQ; prepared under an IPA with Berry College), May 2007. http://mmpp.wikispaces.com/hob30 (accessed May 7, 2008), reproduced by permission.

- When does it occur?
- Where does it take place and what is the defect that resulted?
- Who identified the mistake or defect?
- Who made the mistake?
- Did it happen because of variation of the standard operating process?
- What was the variation that occurred?
- Why was the standard varied?

When performing any analyses, keep from assigning any blame. Simply identify "what happened" as fact rather than put any interpretation to the events or activities. Using the 5 Whys tool can help uproot the answers to these questions. Although the cause of an error may seem obvious, asking "Why?" five times can help reveal source upon source of the mistake or error. (See the "Tool Connection: 5 Whys" sidebar on the next page.)

5 WHYS

The *5 Whys tool* provides a means of discovering root causes. The technique involves continuing to ask "Why?" five times until a root cause is identified. Use of structured questions, including specific inquiries, facilitates this technique's success. This tool is especially helpful in combination with a 5S event. (See Chapter 2 for more on 5S events.) As an example, a nurse was missing an 8:00 A.M. medication dose for one of her assigned patients. To reveal the root cause of the problem, the Lean team asked the following 5 Whys[5]:

Q1. Why was the 8:00 A.M. dose of antibiotic for Mr. U in Room 303, Bed 1, missing?

A. It was missing when the assigned registered nurse (R.N.) went to get the dose from the cart.

Q2. Why was the antibiotic dose missing from the cart when the R.N. went to get it?

A. When cart exchange occurred, the pharmacy technician brought back to the pharmacy an antibiotic dose labeled for Mr. U as "first dose midnight" and credited it at 8:02 A.M.

Q3. Why did the pharmacy technician bring back to the pharmacy an antibiotic dose labeled for Mr. U as "first dose midnight" and credit it as 8:02 A.M.?

A. Because the dose labeled "12 midnight" was not given; however, one dose was charted as given.

Q4. Why was the dose labeled "12 midnight" not given at 12 midnight?

A. Because the dose labeled "8:00 A.M." had been given at midnight, and the date and time of the dose was changed to "12 midnight."

Q5. Why was the 8:00 A.M. dose given at 12 midnight?

A. Because the first dose was not where first doses should be kept and the pharmacy technician found the first dose in an incorrect location.

During the analyze phase, don't jump to conclusions. Keep seeking roots and the roots of roots. Any information that pertains to the mistake should be collected and analyzed. Identify the "red flag" conditions, which include repetition or routine, processes that are only seldom conducted, making faulty assumptions (such as similarity of patient names and drug names), and relying on memory to execute a complicated process. Other red flag conditions include (but are not limited to) the following:

- High patient or client volume
- Poorly defined standards
- Poorly defined sequences of processes
- Poorly designed devices, equipment, supplies, and packaging
- Computerized alerts that can (and often do) get overridden by busy providers
- Lack of methods to confirm accurate information
- Ambiguous inventory systems where location, structure, or setup lead to confusion
- Reliance on illegible handwriting and medical abbreviations in physician orders
- Situations in which providers can be interrupted or distracted when administering medications or caring for patients
- Poor communications between providers at shift transitions and from hospital physicians to primary (outpatient) caregivers
- Lack of confirmation systems when using look-alike/sound-alike medications
- Lack of training for nonproviders who interact with patients
- Reliance on word of mouth of the patient or client in regard to providing medication and other health care history
- Environmentally challenging conditions such as temperature fluctuations or lighting defects

Don't make assumptions about the validity of the information you are preparing for analysis.

Tool Connection

CAUSE AND EFFECT
USING A FISHBONE
(OR *ISHIKAWA*) DIAGRAM

The *fishbone diagram* serves as a visual representation to clearly display the various factors affecting a process (as listed in Figure 4-1). It can be a structured approach to root cause analysis. The diagram identifies the inputs or potential causes of a single output or effect. In a hospital, for example, work can be divided into categories: responding to instructions (orders), using supplies and medications (materials), using equipment (machinery), providing the needed care or service in accordance with established procedures (methods), and the environment itself (environment). Those categories can be listed on a cause-and-effect diagram as the branches from which mistakes can arise. To complete the branches, brainstorm primary causes, ask together why they are occurring, analyze the causes, and prioritize and identify the likely root causes.

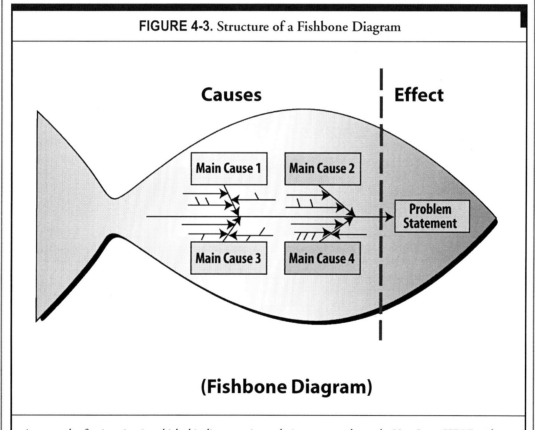

FIGURE 4-3. Structure of a Fishbone Diagram

(Fishbone Diagram)

An example of a situation in which this diagramming technique was used was the New Jersey HIV Family Centered Care Network investigation to identify and address causes of low cervical cancer screening rates.

Source: Armas L., DeLorenzo L., Norberg A., and the Cervical Cancer Screening Subgroup of the AETC Women's Health and Wellness Workgroup: Using a fishbone diagram to assess and remedy barriers to cervical cancer screening in your healthcare setting. AIDS Education and Training Center slide presentation. Presented as a distant-learning program in a Training Exchange through the AIDS Education and Training Centers National Resource Center, Newark, NJ, Oct. 2007. http://www.aids-ed.org/aidsetc?page=etres-display&resource=etres-347 (accessed May 7, 2008).

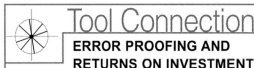

Tool Connection
ERROR PROOFING AND RETURNS ON INVESTMENT

It is often difficult to project what actions will result in a good return on investment (ROI). In his book prepared for the Agency for Healthcare Research and Quality,[4] John Grout, Ph.D., cites articles published by K.R. Bhote[6] and H.M. Whited.[7] Bhote wrote that ROIs of 10 to 1, 100 to 1, and even 1,000 to 1 are possible, but the awareness of error proofing is as low as 10%, and implementation is "dismal" at 1% or less. Whited wrote of exceedingly high rates of return and provided a few examples.

- One corporation employed a device that eliminated a mode of defect that had cost the company $0.5 million a year. The device, which cost only $6, had been created by a production worker in his garage and had a rate of return of 83,333 to 1 for the first year.

- A worker at Johnson & Johnson's Ortho-Clinical Diagnostics Division devised a way to use commercial stickers to reduce defects and save time, the combination valued at $75,000 per year. If the stickers cost $100 per year, the return on investment would be 750 to 1. This example and the previous one represent the return on the use of only one device (solution) each.

- Lucent Technologies' Power System Division implemented the use of 3,300 devices over three years. Each of these devices contributed a net savings of approximately $2,545 to the company's bottom line, and the median cost of each device was approximately $100.

Information must be current, accurate, and verified. Current means as up-to-date as possible. Be precise about the data that are being collected and analyzed or the solutions may not be accurate.

Standardize

This step is for generating ideas. Create standard work instructions and communicate them to staff members who do not follow them. Such communication can be handled through a document that is already in existence. Obtain consensus on any new standard work procedures.

Cost allotments should be judged on return on investment. If errors and defects can be eliminated without a major cost, go that route. However, keeping the patient as the focus, don't avoid cost that could provide the best outcomes—*safety and quality come first.*

Alert

Communicate to staff the conditions that may cause an error. Create "alert" conditions, instituting *visual controls* as much as possible to remind staff of risky red flag conditions.

Error Proof

Create error-proof devices or systems. These may be as simple as standard instructions and forms. Implement the techniques. Create a system to follow up with noncompliant staff.

This last step is the time when the team identifies and implements the level of protection that is most warranted and cost effective. Solutions, devices, and methods are put into place. This is often a difficult step to accomplish because of differences in priorities and contingencies at any given time. Each priority should be followed up by the team leader so that momentum will not be lost. Management should also work to help the team receive the resources it will need to effectively put this step into practice. The list should be posted where everyone can see continual progress improvement. A visual chart to provide at-a-glance progress is helpful to keep on track.

A patient safety alert (PSA) system is a means to implement error proofing. The PSA system involves calling out anything that may harm a patient before injury is incurred. (See Sidebar 4-1 on the next page.)

Techniques for Error Proofing

Self-inspection is an important means of preventing errors. Quality must be built into the original activity, not inspected for upon completion. Outside audits require inspection but are only about 80% effective.[8]

Create policies and standard work procedures that allow staff members to do things right the first time, every time, and catch mistakes before they create harm. Making this happen requires four things[9]:

1. Training on how to properly perform the procedure
2. Process design and standards to meet
3. Equipment or other means to make the inspection. Examples include automated alerts, forced functions (whereby equipment will not operate unless input is accurate), computerized physician order entry (CPOE), and bar coding.
4. Time to perform the procedure properly and to inspect where needed

Building quality and self-inspection techniques ensures that no errors are passed along to the next step because the effort required to correct an error increases dramatically at each successive process step. An illustration of self-inspection and checks appears in Figure 4-4 on the next page.

Medical device, medical equipment, and pharmaceutical manufacturers are following suit in reducing the chance of drug mix-ups by improving clearer differentiation of drugs. For example, there are 19 categories of high-alert medications; studies show that about eight medications, including heparin, account for 31% of all medication errors that harm patients. Hospital staffers sometimes confused similar packaging used for the adult and pediatric versions of the drug. The manufacturer has responded by repackaging high-dose heparin more distinctively to reduce errors.

SIDEBAR 4-1.
Patient Safety Alerts:
One Institution's Experience

A patient safety alert (PSA) system is an important and necessary system in health care. At Virginia Mason Medical Center (VM) in Seattle, a PSA system was implemented in 2002. Any employee can "stop the line" until the problem is resolved. Important components of this system include an institutionwide policy statement, the commitment of senior executives, dedicated resources, a 24-hour hotline, and an outline for communication procedures.[10]

Although initially staff members were slow to call PSAs, as awareness of the policy grew and as improvements were incorporated into systems and processes, the number of PSAs increased dramatically. Staff shared with administration that when they called out a PSA and executives deemed that it was not truly significant enough to be called a PSA, staff felt unheard and ignored.[10,11] Consequently, VM leadership learned that terming an event "significant" was in the eye of the beholder. In addition, administration also realized that staff often felt confused about whether an event pertained to quality versus safety, and what differences that made. The original policy required any staff member who encounters a situation that he or she believes is likely to harm a patient to use the 24-hour hotline to immediately report to the department of patient safety. The staff member also must stop any activity that he or she believes could cause further harm. The resulting revised policy statement, issued in February 2005, differentiated between high, moderate, and low likelihood of recurrence.

For the PSA initiative at VM, a new role of patient safety specialist was created, and claims managers and quality resources specialists (usually nurses) were invited to apply for that position. A Web-based reporting system also was begun so that staff could report PSAs 24/7. When a report is entered, the on-call patient safety specialist is alerted, and the case is entered into the database, which is subject to analysis and aggregate reporting (see Figure 4-5 on the next page).

FIGURE 4-4. Self-Inspection Helps Prevent Error

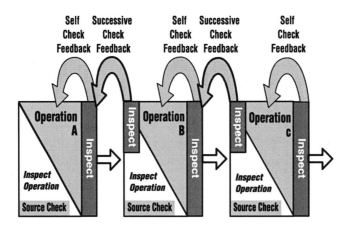

Requiring self-checks and successive checks along a process path helps prevent errors.

Source: Lowstuter B., Hagood C.: Healthcare Performance Partners: Mistake-proofing through source checks, self checks and successive checks. *Lean Healthcare Exchange,* Mar. 12, 2008. http://leanhealthcareexchange.com/?p=102 (accessed Jul. 7, 2008), reproduced by permission.

FIGURE 4-5. Patient Safety Alerts

Categories of Patient Safety Alerts (PSAs), 2002–2006					
Category*	2002 (5 mos)	2003	2004	2005	2006
Adverse drug reaction				23	31
Airway management				24	28
Blood/blood component				29	16
Care/service coordination				452	1103
Diagnosis/treatment	6	6	19	62	61
Diagnostic test				82	100
Employee general incident				35	27
Environment (facilities)	1	22	22	73	76
Fall				260	285
ID/documentation/consent				166	170
Infection control				116	124
Lab specimen/test				449	405
Line/tube				16	22
Medication/IV safety	4	17	32	448	580
Restraints/supportive devices				6	5
Safety/security/conduct	3	6	13	78	119
Skin/tissue				20	34
Surgery/procedure				84	99
Surgical site infection				11	2
Systems†	4	74	118		
Vascular access device				16	28
Total declared PSAs	18	125	204	2450	3315

* ID, identification; IV, intravenous.
† Since 2004 "Systems"-related PSAs have been reassigned to more specific categories.

This figure shows one format for reporting patient safety alerts.

Source: Furman C., Caplan R.: Applying the Toyota Production System: Using a patient safety alert system to reduce error. *Jt Comm J Qual Patient Saf* 33:376–386, Jul. 2007.

Tool Tangent

AUTONOMATION

Autonomation (preautomation) was originally a feature of machine design to effect the principle of "stopping the line" used in the Toyota Production System. Applying autonomation "with a human touch" is an excellent means of error proofing.[9] A machine will sense an error and shut down automatically or call for operator intervention to correct the problem. An example is the case of a pulse oximeter becoming loose or disconnecting from the patient so that the reading deviates from the expected range. This deviation will set off an alarm, which requires human intervention to silence the signal and to replace the sensor on the patient's finger. Autonomation follows four basic principles:

1. Detect the abnormality.
2. Stop the process ("stop the line").
3. Fix or correct the immediate condition.
4. Investigate the root cause and install a countermeasure.

An example of this concept is the common practice of blood banks to group several test tubes in one rubber band with a slip of paper identifying the blood. But the slip of paper would sometimes fall out, creating the opportunity for the tubes to be misidentified. The tubes could also shift, occasionally falling and breaking, creating a biohazard. The industry has responded to this risk by developing a plastic rack that securely holds the test tubes and accompanying information; in some cases, preprinted device serial numbers can be used to improve traceability. Another example is the automatic alert shown in Figure 4-6.

FIGURE 4-6. Preventive Mechanisms

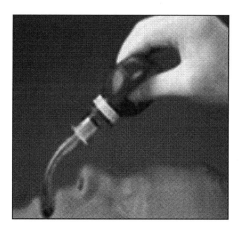 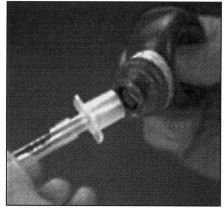

Putting a pulmonary tube into a patient's stomach is prevented by this mechanism. Solution: Squeeze the plastic bulb and put it on the tube. If bulb inflates, the tube is in the lungs. If not, an error has occurred.

Source: i Six Sigma Healthcare: *John Grout's Mistake-Proofing Center,* example 23. 2008. http://healthcare.isixsigma.com/offsite.asp?A=Fr&Url=http://www.mistakeproofing.com (accessed Mar. 21, 2008), reproduced by permission.

Tool Connection
ROOT CAUSE ANALYSIS

Root cause analysis (RCA) is a set of methodologies used to determine one or more root causes of an event. The technique attempts to root out the core origins of problems that can be controlled or modified so that the event will not occur again given the same set of circumstances. RCA methodologies reveal the cause-and-effect relationships that exist in a system. RCA is best used in assessing rare events, such as wrong-surgery or egregious medication errors, as opposed to common patient safety problems, such as hospital-acquired infection or contrast-induced nephropathy. The five steps of RCA include the following:

1. Data collection
2. Causal factor charting
3. Root cause identification
4. Generating recommendations
5. Implementing recommendations

When using RCA, caution should be applied to formulate corrective actions because the process may consider only one instance or circumstance of failure.

Levels of the Process Sequence

Errors can occur at different times during the period of a process's duration. Attention at all levels ensures zero defects. There are three levels at which error proofing can occur[12]:

1. Level 1 eliminates error at the source before it occurs.
2. Level 2 detects the error as it occurs and before it results in a defect.
3. Level 3 detects the defect after it has been made but before it reaches the next process or department.

For instance, Level 1, which is obviously the safest level at which to catch a potential error or defect, would include a visual control for triple-checking medication dispensing and perhaps a checklist or computer alert. An example in hospitals is an allergy alert appearing in red text on a patient's chart and identification (ID) bracelet.

A Level 2 visual control would be the computerized alert that occurs when typing the patient's medications into a computer software file. This type of visual control would alert the health care provider to a drug interaction.

A Level 3 error-proofing visual control would be discovering during a preanesthesia procedure for a right femur tumor removal that hair had been removed from the left leg during the preoperative process; the defect is having prepared the incorrect surgical site, but the error has been caught in time so that the patient has not yet been harmed.

Relevant Health Care Application

Medication errors are a huge problem in health care and represent an area that is being targeted nationwide in the United States. The VA Medical Center, which is part of the Pittsburgh Regional Health Initiative, targeted for a project the use of gloves in seven beds on one unit in order to reduce the rates of methicillin-resistant Staphylococcus aureus (MRSA) infections.[13,14] Interventions included glove dispensers mounted on the wall in every patient room and visual indicators reminding everyone, from housekeeping to providers and assistants, to wash their hands before and after they touched patients. The system included a non-soap-based cleaner in dispensers, and red lines were drawn around beds to indicate that when anyone crossed the line, hand washing was needed. Compliance increased and the rates of MRSA decreased.

REFERENCES

1. Spear S.J.: Fixing health care from the inside, today. *Harv Bus Rev* 83(9):78–91, 158, 2005.

2. Printezis A., Gopalakrishnan M.: Current pulse: Can a production system reduce medical errors in health care? *Qual Manag Health Care* 16(3):226–238, 2007.

3. Simon K.: *Poka yoke mistake proofing.* i Six Sigma Healthcare, 2008. http://healthcare.isixsigma.com/library/content/c020128a.asp (accessed Mar. 21, 2008).

4. Grout J.: *Mistake-Proofing the Design of Health Care Processes.* AHRQ Publication No. 07-0020. Agency for Healthcare Research and Quality (AHRQ; prepared under an IPA with Berry College), May 2007. http://mmpp.wikispaces.com/hob30 (accessed May 7, 2008).

5. Thompson D.N., Wolf G.A., Spear S.J.: Driving improvement in patient care: Lessons from Toyota. *J Nurs Adm* 33:585–595, Nov. 2003.

6. Bhote K.R.: A powerful new tool kit for the 21st century. *National Productivity Review* 16(4):29, Autumn 1997.

7. Whited H.M.: *Poka-Yoke Varieties. The Power of Mistake Proofing Forum,* August 6, 1997. Moline, IL: Productivity, Inc., Autumn 1997.

8. Lowstuter B., Hagood C.: Healthcare Performance Partners: Mistake-proofing through source checks, self checks and successive checks. *Lean Healthcare Exchange,* Mar. 12, 2008. http://www.leanhealthcareexchange.com/?p=102 (accessed Jul. 7, 2008).

9. Manos A., Sattler M., Alukal G.: Make healthcare Lean. *Quality Progress,* Jul. 2006. http://www.asq.org/qualityprogress/past-issues/index.html?fromYYYY=2006&fromMM= 07&index=1 (accessed Mar. 21, 2008).

10. Furman C., Caplan R.: Applying the Toyota Production System: Using a patient safety alert system to reduce error. *Jt Comm J Qual Patient Saf* 33:376–386, Jul. 2007.

11. Furman C.: Implementing a patient safety alert system. *Nurs Econ* 23:42–45, Jan.–Feb. 2005.

12. Hadfield D., Holmes S.: *The Lean Healthcare Pocket Guide: Tools for the Elimination of Waste in Hospitals, Clinics and Other Healthcare Facilities.* Chelsea, MI: MCS Media, 2006.

13. Pittsburgh Regional Healthcare Initiative puts new spin on improving healthcare quality. *Qual Lett Healthc Lead* 14:2–11, 1, Nov. 2002.

14. Feinstein K.W., Grunden N., Harrison E.I.: A region addresses patient safety. *Am J Infect Control* 30(4):248–251, 2002.

CHAPTER FIVE

SIX SIGMA

Six Sigma At a Glance

TOOL DESCRIPTION	Six Sigma is a measure of quality control that signifies a process or system is near perfect, that is, it is defined as within six standard deviations from the norm.
TOOL PURPOSE	Improve business performance, by controlling and understanding variation, to ensure predictability of processes
WHO IS RESPONSIBLE?	Facilitated by a Lean Black Belt–trained staff member in cooperation with a cross-functional team
HOW LONG DOES IT TAKE?	1 to 18 months, depending on the complexity of the problem
STEPS IN THE PROCESS	1. Define.
	2. Measure.
	3. Analyze.
	4. Improve.
	5. Control.

Six Sigma is a problem-solving, process-innovation approach for improving business performance. It works by helping staff understand and control variation to improve process predictability. Six Sigma uses data and rigorous statistical analysis to identify "defects" in a process, service, or product; reduce variability; and achieve as close to zero defects as possible. The approach has grown from the synthesis of a series of developments in quality improvement that date to the early part of the 20th century. The term *Six Sigma* derives from a statistical definition of "near perfect," or six standard deviations from the mean value of a process output or task (1 error in 300,000 operations).

Certification for Green Belts (employees who are interested in Six Sigma), Black Belts (project leaders), and Master Black Belts (quality leaders and managers) is available from various consulting organizations, and qualifying criteria may vary among them. The Six Sigma process should be facilitated by a Black Belt–trained staff member, that is, someone who has been certified as having successfully completed an improvement activity with a defined cost savings.

Although in the past Lean and Six Sigma were believed to be opposing theories, it is now well accepted that the two methodologies can work well under the same philosophy, and the synthesis of the two is increasingly being called "Lean Sigma." (For a comparison of the two, see Table 5-1.) In reality, Six Sigma can be thought of as the ultimate Lean process, requiring mastery of standardization, error proofing, value stream mapping, root cause analysis, *kaizen*, and 5S, to name just a few relevant Lean tools.

Logically, because of the sophistication of Six Sigma, it is usually recommended that simpler Lean tools and principles be applied first to eliminate non-value-added activities and to create flow. Although Six Sigma is widely considered an advanced tool that should be used only after implementing other Lean tools first, the personal preference of the Six Sigma Black Belt leading the team may dictate otherwise.

	Lean	Six Sigma
TABLE 5-1. Lean and Six Sigma: Similarities and Differences		
Description	An integrated system of principles, practices, tools, and techniques focusing on standardizing solutions to common organizational problems by reducing waste, increasing value, synchronizing work flow, and managing variability in production flow	Data-driven approach to quality improvement that uses statistical analysis to reduce or eliminate process variation
Purpose	Lean experts use statistical or graphical techniques to eliminate defects and address process velocity (time or speed).	Six Sigma experts address process stability.
Benefit	Most appropriate for reducing cost and time to benefit throughput and productivity	More directly helps maintain and improve quality through reduction in variation
Focus	Customer	Customer
Goal	Zero waste	Zero defects
Aim	Prevent defects	Prevent errors
Teamwork	Lean team works intimately with all internal disciplines as well as external suppliers and customers across an entire organization.	Core group of employees is formally trained in this tool for specific project initiatives.

A Lean Sigma framework specifies exactly how work is to be conducted, to a level of detail such that even small deviations from expected outcomes are evident. Once these deviations are detected, by anyone, they can be investigated and contained. Data gathered in the analysis of process flow, such as through value stream mapping (as described in Chapter 1), can later be used to identify the highest priorities or areas of highest impact for Six Sigma projects.

Causes of Variation

W. Edwards Deming referred to the two types of variations possible as "special cause" and "common cause."[1] Special cause variation results from an arbitrary occurrence at some point in a process. Common cause variation refers to the inconsistencies that are always present, to some

Tool Connection

THE SIX SIGMA EQUATION

The Greek letter *sigma* ( ) is used to denote the sum of a series of numbers. The lowercase *sigma* ( ) denotes the standard deviation, which is generally defined as the average distance from the average value of a set of numbers. Standard deviation indicates whether the data are clustered together or spread out. Six Sigma is defined as plus or minus six standard deviations from the mean. The farther away the spread is from the mean, the lower the quality.

Six Sigma's rationale indicates that the output of a process is a function of all its inputs. This is seen as an equation:

$$y = f(x)$$

In the equation, *y* signifies the output of a process, *x* signifies the input, and the letter *f* stands for a "function of." Six standard deviations, or *Six Sigma*, comprise 99.99966% of the data under a histogram's curve. This translates to 3.4 defects produced per 1 million opportunities (see Figure 5-1). The goal of Six Sigma is to have no more than that proportion of defects. In 1999, the Institute of Medicine's landmark study, *To Err Is Human*,[2] estimated that in hospital care alone, there were about 2.9 defects per 100 opportunities.[3]

Histograms are useful as related Six Sigma tools because they show the distribution away from the mean. An example histogram is shown in Figure 5-2.

FIGURE 5-1. The Six Sigma Approach to Zero Defects

What is Six Sigma?

Sigma	Defects per Million opportunities
2	308,537 (69.1% good)
3	66,807 (93.3% good)
4	6,210 (99.4% good)
5	233 (99.98% good)
6	3.4 (99.99966% good)

3 to 6 – 20,000 Times Improvement... A True Quantum Leap

By limiting possible variations in a process, perfection becomes increasingly more possible. In essence, this means that through Six Sigma, process variation is reduced to a level at which no more than 3.4 parts per million fall outside the specification limits.

Source: © 2008 Juran Institute, reproduced by permission.

CONTINUED

Tool Connection
THE SIX SIGMA EQUATION
CONTINUED

FIGURE 5-2. Histogram

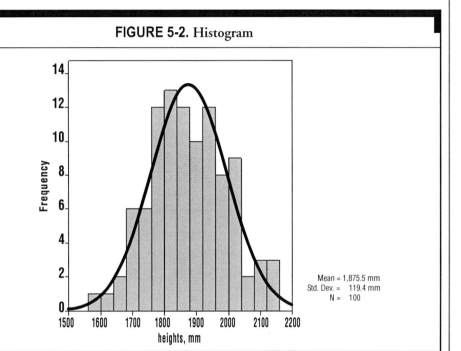

Mean = 1,875.5 mm
Std. Dev. = 119.4 mm
N = 100

The pillars depict the heights of 100 men who have heights between the limits shown on the horizontal scale of heights. The information to the bottom right of the figure gives the mean or average height for these men and the standard deviation. The mean (average) height was 1,875.5mm. The standard deviation is a measure of the spread of the heights in the sample, or 119.4mm.

Source: Mathematics Support Centre: Frequently Asked Questions. Question 12, Coventry University, United Kingdom, Faculty of Engineering and Computing. http://www.coventry.ac.uk/ec/maths_centre/faq/faqSA15.html (accessed May 8, 2008), reproduced by permission.

extent, in every process. Six Sigma is a means of determining the inputs that are critical to quality with the ultimate goal of eliminating or minimizing process variation, no matter what the cause, and thus achieving the process's intended goal. The probable causes of variation can be the actual process itself or the measurement used.[4] To address actual process variability, the variation resulting from the measurement system must be differentiated from that of the process. Measurement variations may be identified as those of accuracy, repeatability, reproducibility, stability, and linearity. (Figure 5-3 on the next page depicts causes of variation.)

Steps in the Process

Because Six Sigma tools are heavily data driven, the necessary infrastructure to collect useful data has not been readily available in health care until recently. The advent of sophisticated technologic software and high-powered computer hardware has made Six Sigma analysis much

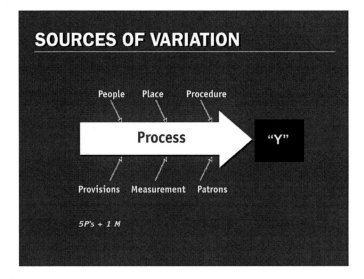

FIGURE 5-3. Causes of Variation

Variation can arise from people, settings, procedures, provisions, metrics, and customers.

Source: © 2008 Juran Institute, reproduced by permission.

easier to do. The absence in many cases of standard operating procedures (standard work) has also made measurement more difficult.

Effective health care delivery calls for the coordination of thousands of "microsystems"—that is, the building blocks that high-performing systems engage: small groups of people performing processes and activities that make up the system whole.[3] Many of these microsystems are repetitive, similar to the operations in the automotive industry, which makes Six Sigma methods useful for extrapolation to health care.

The application of Six Sigma incorporates a methodology that is represented by the acronym DMAIC (pronounced *duh-MAY-ick*). The following are the five phases of Six Sigma implementation:

1. Defining
2. Measuring
3. Analyzing
4. Improving
5. Controlling

In some ways, these steps mimic those of error proofing (as discussed in Chapter 4). As shown in Figure 5-4, DMAIC gives a process team multiple tools that are applicable to specific phases of implementation.

The Wellmont Health System, serving northeastern Tennessee and southwestern Virginia, conducted a Six Sigma project wherein the *Y* factor was the variation in the physician consultation process. The organization's goal was to eliminate the variation associated with delays

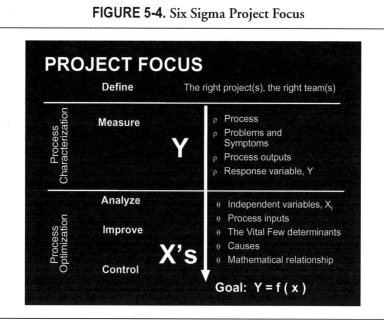

FIGURE 5-4. Six Sigma Project Focus

Six Sigma's project focus identifies X factors (process inputs) and Y factors (process outcomes) to reduce variation.
Source: © 2008 Juran Institute, reproduced by permission.

and/or the failure to provide a consultation to a patient as needed. Wellmont's experience with the steps in this Six Sigma project is explored in the following sections.

Define

Define the customers, their requirements, the team charter, and the key process that affects the customer. Clearly answer the question: What is the problem?

During this step, a project charter is developed. This charter is similar to an event planning worksheet and employs similar guidelines. In this phase, a business case is laid out and timing reviewed. The business case section determines the importance of the project to the organization and how the project is important to the organization's strategic plan. A business case also employs a cost-benefit analysis that identifies expected financial benefits (such as cost reduction or profit increase) and their derivation. The review timing section steadily tracks the project's progress. The project may take months to a year to complete; therefore, periodic reviews help ensure the project's continuing momentum. In this phase, the project is scheduled for a solution and assigned to a team headed by a Six Sigma Green Belt or Black Belt who will report to a leadership champion within the organization. If it is assigned to a Green Belt, that person should have a Black Belt available to assist when needed and should be completing course work to improve his or her Six Sigma skills.

In the design phase at Wellmont, the team was formed. It gathered information regarding the baseline for the project. Using its Team Charter Page, it identified the problem statement, objective, historical performance, Y (output), and measures of quality as perceived by the customers (patient, nurse, physician, secretary). The charter page serves as the team contract. Because historical information was somewhat incomplete, the team relied on incident reports and surveys filled out by key staff members who reported many of their struggles with the consultation process.

Measure

Identify the key measure and a data collection plan. Clearly answer the question: What is the extent of the problem?

During this phase, baseline data are collected and a diagnosis is begun by determining a key measure for the project. The data are translated into critical-to-quality characteristics.[5] Data must be collected in such a way as to contribute to the system's validity and reliability. Intended goals also need clarity. Using critical thinking is imperative to make sure that units of data are appropriate, current data are collected, collection methods are as accurate as possible, and the duration of time over which data will be collected is determined before beginning. Data collected during this phase serve as the baseline against which further data will be compared.

In the Measure phase at Wellmont, the team used tools such as a detailed process mapping, cause-and-effect diagrams, and a data collection plan. The detailed process map was used to identify the process steps and their necessary inputs to each step. Each input was categorized in one of the following ways:

- Controllable (the person can make changes as necessary to achieve the desired output)
- Standard operating procedure (SOP)
- Ideal type of input (the input variable X has a documented procedure of how it is to be completed)
- Noise (the person has little control on the input, which is random)

Cause-and-effect diagrams were used to identify a complete list of possible sources of variation in the process. As the project is still in process, the team will use this document to prioritize the processes that need improvement first. In addition, a data collection plan will be designed to capture delays and omissions.

Analyze

To continue the Six Sigma event, analyze the data and the process to determine the root cause(s) for why the process is not performing as desired. Clearly answer the question: Under what circumstances does the problem occur?

In this phase, statistical analysis is used to calculate the probability that an end result will occur. In Six Sigma, statistical analysis is used to demonstrate how the process inputs impact on the process outputs (as cause and effect). An expert in statistical analysis will need to lead the Analyze phase and employ applicable tools. (For an example analysis tool, see Figure 5-5.) An analysis of the data collected in the Measure phase has determined the probability of obstacles or barriers occurring during the process and the origins of those problems. Most important, the causes that are identified during this phase will serve as the foundation for the next phase in the process. This is where the real value of the Six Sigma methodology shines.

In the Analyze phase at Wellmont, once the list of Potential Xs (from the cause-and-effect diagrams) has been prioritized, the team will work to identify them as Key Process Input Variables

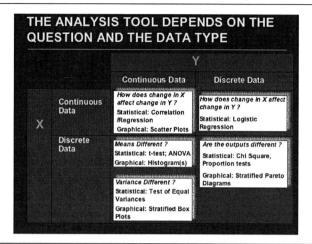

FIGURE 5-5. Determine the Question and Data Type

THE ANALYSIS TOOL DEPENDS ON THE QUESTION AND THE DATA TYPE

		Y	
		Continuous Data	Discrete Data
X	Continuous Data	*How does change in X affect change in Y ?* Statistical: Correlation /Regression Graphical: Scatter Plots	*How does change in X affect change in Y ?* Statistical: Logistic Regression
	Discrete Data	*Means Different ?* Statistical: t-test; ANOVA Graphical: Histogram(s) *Variance Different ?* Statistical: Test of Equal Variances Graphical: Stratified Box Plots	*Are the outputs different ?* Statistical: Chi Square, Proportion tests Graphical: Stratified Pareto Diagrams

The analysis tool depends on the question being answered and the data type. X represents process inputs; Y, process outputs.

Source: © 2008 Juran Institute, reproduced by permission.

(KPIVs). Some of the KPIVs will entail immediate changes to the process, and others will need to be improved based on changing inputs from controllable to SOP. The team will also use hypothesis testing methods such as correlation and regression to determine the relationship between the secretary, physician, time of day, day of week, and so forth. The goal is to develop a process that will prevent these relationships from affecting organizational performance. Using other tools—such as failure mode effects analysis (FMEA), a technique for analyzing the risk associated with potential problems—the team will explore the overall consult system to understand how each process may fail and improvement efforts will be prioritized based on the risk priority number (RPN).

Improve

Generate potential solutions and implement them on a small scale to determine whether they positively improve the process performance. Clearly answer the question: How can the problem be solved?

In this phase, the solutions to the identified problems are developed, evaluated, refined, and implemented. Experimental options are designed to determine the value of each solution toward the end goal. Results of trial and error will indicate which solutions should be verified before they are finally implemented. Making small-scale improvements helps adjust and adapt conditions that must be modified.

Process improvement using Lean and Six Sigma might first appear to initiate a decline in quality due to increased event reporting. In fact, this reporting does not designate a failure but, on the contrary, may show a success in the project's interventions and staff willingness and dedication to Lean/Six Sigma thinking.

During the Improve phase, the Wellmont team introduces the new process to staff and implements a system to monitor the performance by using a process capability study, as shown in

---

FIGURE 5-6. Process Capability Analysis

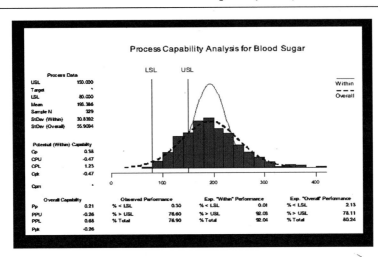

Six Sigma seeks to improve process capability by recognizing ambiguity, reducing variability and errors in care delivery, and improving operational performance. This example of the glucose levels of diabetic cardiac surgery patients demonstrates process capability, where the dashed lines indicate overall rates and the curved lines indicate those falling within six standard deviations (within six sigma).

Source: Slide presentation given by Joseph M. Duhig, Senior Vice President, Juran Institute and Juran Healthcare; Six Sigma Master Black Belt; and former CEO and Executive Director, University Medical Center Alliance, Memphis; to the Advanced Training Program of the Intermountain Institute for Healthcare Delivery Research in Salt Lake City, Feb. 13, 2008. http:// intermountainhealthcare.org/xp/public/institute/library/faculty16.xml (accessed Jul. 8, 2008), reproduced by permission.

Figure 5-6. This study will tell the team how it is performing based on a target goal. It will provide short-term and long-term capability and expected performance. The team will train personnel based on this new process and will provide them with a physician consultation matrix that contains the necessary information they need to determine procedures such as whom to call, how to contact, and when to contact. The project leaders also expect to continue using FMEA as they continue to make changes to the process. They also will update the detailed process map, citing any changes and looking for data collection opportunities that may help them streamline the operation in the future.

Control

Develop, document, and implement a plan to ensure that performance improvement remains at the desired level. Clearly answer the question: How can we ensure that this solution endures?

In this final phase of Six Sigma, improvements and design controls are documented. Changes and improvements will need to be maintained. In essence, control means that standard work is developed and adhered to and integrated within the overall system.

In the Control phase at Wellmont, the team will use such tools as measurement system analysis (MSA) to evaluate an individual's ability to follow the process flow and determine the appropriate physician to contact. The team will restate the capability of its process based on these new guidelines, once again looking for performance based on a target. In addition, the

---

TABLE 5-2. Various Uses of Six Sigma in Health Care

Clinical examples	Nonclinical examples
• Improve lab work flow and test turn-around time. • Reduce emergency department (ED) wait time and improve patient satisfaction. • Reduce days patients are on ventilators and intensive care unit (ICU) length of stay; reduce pressure ulcers. • Increase on-time starts for first-case surgeries. • Increase inpatient bed availability to reduce ED diverts. • Shorten length of stay in chronic obstructive pulmonary disease (COPD) patients. • Reduce number of patients who need intravenous antibiotics. • Shorten preparation time of intravenous medications.	• Compare different physicians' costs for treating common chronic diseases, such as diabetes. • Help practices determine what percentage of patients has a specific disease. • Improve patient education, thereby reducing individual patient visits. • Improve the claims handling process, including electronic claims cycle time and accuracy, so that payments can be made more quickly. • Ensure that specifications, modules, and resource assignments align efficiently when launching software development work. • Reduce printing costs. • Reduce hospital employee turnover in a hospital system. • Reduce errors in invoices received from temporary employment agencies. • Revise terms of payment. • Allow parents to "room in" with their hospitalized children. • Reduce the number of mistakes in invoices.

team will implement a control plan that provides information as a written summary describing the systems that will be used to control product and process variation to an acceptable level. The control plan is a "living" document that can and should be updated periodically. It also provides information on how the process will be audited for performance.

Relevant Health Care Applications

Many health care professionals believe that quality improvement methods should be used only to address defects in systems, such as medication errors. However, those experienced with Lean and Six Sigma think extending the applications in which Six Sigma is used can reduce operational inefficiencies across the board.[5] Health care institutions have used Lean/Six Sigma in many ways.[1,5,6,7,8] (See Table 5-2.)

Tool Connection

RELATED TOOLS

Six Sigma makes use of a great number of established quality management methods that are also used outside of Six Sigma. These tools include the following:

- *Run charts:* data graphs run over time. These charts are important, frequently used performance improvement tools. The benefits of run charts include depicting how well a process is performing, displaying a pattern of data that can be observed during changes, and providing direction about the value of particular changes.

- *Affinity diagram:* Used after brainstorming, the affinity diagram helps organize ideas so that connections between issues are logical.

- *Scatter diagram:* a visual representation of data used to study the possible relationship between one variable and another. The scatter diagram graphs pairs of numerical data; one variable is charted on each axis. If the variables are correlated, the points will fall along a line or curve; the better the correlation, the more tightly the points will adhere to the line. This tool is used when numerical data are being paired, when the dependent variable has multiple values for each value of the independent variable, or when trying to determine whether the two variables are related. Examples of this use are when identifying potential root causes, after brainstorming causes and effects using a fishbone diagram, when determining whether

FIGURE 5-7. Scatter Diagram

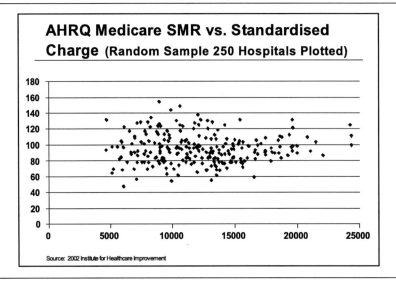

This example of a scatter diagram plots reimbursement versus mortality rate.

Source: Slide presentation given by Joseph M. Duhig, Senior Vice President, Juran Institute and Juran Healthcare; Six Sigma Master Black Belt; and former CEO and Executive Director, University Medical Center Alliance, Memphis; to the Advanced Training Program of the Intermountain Institute for Healthcare Delivery Research in Salt Lake City, Feb. 13, 2008. http://intermountainhealthcare.org/xp/public/institute/library/faculty16.xml (accessed Jul. 8, 2008). Reprinted with permission of the Institute for Healthcare Improvement (www.IHI.org).

CONTINUED

Tool Connection

RELATED TOOLS
CONTINUED

two effects that appear to be associated occur with the same cause, and when testing for autocorrelation prior to building a control chart (the visual representation of tracking progress over time). Figure 5-7 on the opposite page shows an example of a scatter diagram.

• *Paynter chart:* Based on the Pareto principle, this chart focuses on the areas of priority but graphically displays data by subgroups. An example is presented in Figure 5-8 below.

• *Process mapping:* the visual representation of a sequence of tasks consisting of people, work duties, and transactions that occur for the design and delivery of a product or service

• *Team charter:* a document detailing the team's mission and proposed outcomes to ensure strategic alignment. A team charter typically includes the categories of project scope, leadership, schedule, current state issues, goals and objectives, team members, potential deliverables, and possible obstacles to success.

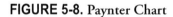

FIGURE 5-8. Paynter Chart

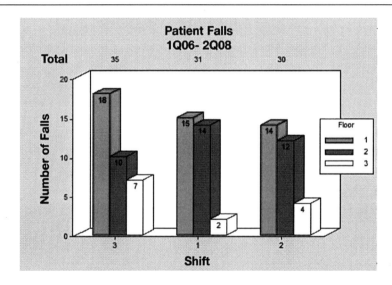

This is an example of a Paynter chart with the graphic representation of subgroups. This tool, which was developed at the Ford Motor Company, combines the concepts of a run chart with a Pareto chart.

Source: Statit Solutions Group, © 2008, reproduced by permission.

In health care, Six Sigma analysis has been used primarily in nonclinical areas, yet an example of using Six Sigma to uncover inappropriate treatment is with anemia. Anemia is a common blood disorder stemming from the reduction in the number of healthy red blood cells. Patients who are anemic suffer from debilitating effects including blackouts, fatigue, and low quality of life. Using Six Sigma tools, researchers and a Six Sigma expert sought repeatability and reproducibility in the measurement of red blood cell count.[9] The high-level X factors for anemia included increased destruction of red blood cells, increased blood loss from the body, and inadequate red blood cell production by the bone marrow. Pulling sample data, this combination accounted for anemia in more than 1,373 patients over one year's time. Using a multivariate (that is, using multiple independent variables), passive Six Sigma tool on these data, analysts sorted by cause, codes, and diagnostic conditions to separate out 286 active patients for the study. Of those active patients, 23% ($n = 65$) had a type of anemia caused by blood loss, so that was chosen as the Six Sigma *Y* factor. The physicians had treated the patients' symptoms of anemia for years but had never investigated the root causes.

Within one month of analysis, it was demonstrated that for the 17% of the 65 active patients (12 patients), the root cause was the passage of blood through stool. With the performance of a surgical procedure to correct that "defect," those patients are now living symptom free. Overall, the surgical solution cost $7,000 per patient, compared to an average cost of $78,000 over a two-year span of the prior treatment protocol.

Reducing Variation in Heart Attack Treatment

At Wellmont Health System, variation is the target for heart attack treatment.[7] Average door-to-balloon (D2B) times for heart attack patients now equal less than 40 minutes. Current guidelines call for D2B to be accomplished within 90 minutes and use a cardiovascular template that includes aspirin, ACE inhibitors, and beta-blockers, as is the recommended treatment in all hospital settings. In addition, patients arriving at Wellmont's emergency department (ED) who are 16 years old or older with unspecified chest pain are placed in a room within 2 minutes of triage. They receive an electrocardiogram (EKG) within 4 minutes, and the EKG is read within 5 minutes. If the EKG is positive for the acute coronary syndrome ST-segment elevation myocardial infarction, a call is made to the catheterization lab, and the patient is admitted to the hospital within 15 minutes. After medications are administered, the patient is released from the ED within 25 minutes.

Building Safeguards to Reduce Patient Falls

Another project in the Wellmont Health System, "Catch a Falling Star," targets variation to reduce patient falls.[7] Bundling is an efficient means to implement standard work and Six Sigma attention to variation; this project bundled safeguards in six steps. First, every patient gets a fall assessment. Then, those identified as high risk receive a bed alarm, a star on their charts, a sign on the door, hourly rounding, and knee-length pajama bottoms to prevent tripping. In addition, careful attention to the floor-washing schedule prevents patients' slipping on wet floors. As of March 2008, fall rates had decreased from five falls per 1,000 days to one fall per 1,000 days, and the compliance rate was approaching the 95% mark. Wellmont will continue the emphasis on achieving 100% compliance and zero defects (falls).

Pilot Testing to Improve Medication Administration

At Deaconess Glover Hospital in Boston, a senior surgeon and several senior administrators initiated a pilot test to address variations in medication administration.[2] Problems included nurse interruptions and variations in the methods for making medication requests, including multiple forms. To reduce the ambiguity and variation, the hospital used standardization of work, simplification and automation, and pathway redesign. Errors in medication administration were drastically reduced.

Using Six Sigma to Streamline Blood Draws

At DSI laboratories in Naples, Florida,[10] in order to have all test results on patient charts by the required time of 7:00 A.M., phlebotomists had to draw blood samples between 3:00 A.M. and 5:00 A.M. Because all 12 phlebotomists arrived at the lab at about the same time, a bottleneck of specimens overwhelmed the laboratory staff. Using Lean and Six Sigma methods, the team created a smoother work flow by devising a system whereby blood was drawn beginning the night before, from 9:00 P.M. to midnight. Drawing would resume at 3:00 A.M. for patients in the ICU, and blood samples from any remaining patients would be drawn from 4:00 A.M. to 6:00 A.M. After 10–15 patients, phlebotomists brought their samples to the lab and returned once more to the floors. Tubes now arrived to the lab at an even pace during the day, and the needed phlebotomy staff at the 3:00 A.M. draw was reduced from 12 people to 2. In addition, a 5S project standardized cart labeling, processing manuals, and bulletin boards. The overall lab savings from these efforts reduced overtime spending by 60% to $78,000 in just the first year. Six phlebotomists were reassigned to open positions, saving $160,000 annually. The Lean project also required 4.5 fewer technologist positions, which saved $250,000. In total, the savings exceeded $400,000 just in the first year.

Decreasing Variation to Reduce Medication Errors

Medco Health Solutions, Inc., Franklin Lakes, New Jersey,[11] provides pharmacy benefit management services to moderate the cost of prescription drugs and to enhance the quality of pharmaceutical care. One of the company offerings is a mail-service delivery of prescription medications through a home-delivery pharmacy. Medco implemented a number of Six Sigma activities to boost alerts and system enhancements for look-alike/sound-alike (LASA) drugs. The organization's entire team underwent extensive Green Belt or Black Belt training. To reduce the significant process variation identified, the team did the following:

1. Ensured that the local pharmacy standard operating procedures were consistent across the board and were aligned with corporate standards
2. Reinforced ongoing education, awareness, and training for pharmacists, specifically about commonly occurring medication errors
3. Initiated a procedure for developing, reviewing, and intensifying LASA alerts

From its interventions, the team decreased the number of variations. A linear regression analysis demonstrated reductions in the following categories of medication errors:

- Wrong-drug selection: 33%
- Wrong directions: 49%
- LASA errors: 69%
- Supply errors: 48%
- Patient name errors: 46%

In the early phases of the project, the team faced challenges and had to overcome a skeptical organization view of change management, a manual data collection process, process variation among pharmacies, and the need to restructure the medication error assessment process.

REFERENCES

1. Deming W.E.: *Out of the Crisis.* Boston, MA: MIT Press, 1986.

2. Institute of Medicine: *To Err Is Human: Building a Safer Health System.* Washington, DC: National Academy Press, 1999.

3. Printezis A., Gopalakrishnan M.: Current pulse: Can a production system reduce medical errors in health care? *Qual Manag Health Care* 16(3):226–238, 2007.

4. Duhig J.M.: Six Sigma and Lean production. Slide presentation given to the Advanced Training Program of the Intermountain Institute for Healthcare Delivery Research in Salt Lake City, Feb. 13, 2008. http://intermountainhealthcare.org/xp/public/institute/library/faculty16.xml (accessed Jul. 8, 2008).

5. de Koning H., et al.: Lean Six Sigma in healthcare. *J Healthc Qual* 28(2):4–11, 2006.

6. Hill D.: Physician strives to create lean, clean health care machine: Studies of manufacturing processes may one day help make your practice more efficient. *Physician Exec* 27:62–65, Sep.–Oct. 2001.

7. Lazarus I.R., Andell J.: Providers, payers and IT suppliers learn it pays to get "Lean." *Managed Healthcare Executive* 16(2):34–36, Feb. 2006.

8. Breakthrough Management Group International: Lean Six Sigma case studies. http://www.bmgi.com/success_stories/casestudies.aspx (accessed Mar. 21, 2008).

9. Brue G.: The elephant in the operating room. *Quality Digest* 25(6):49–55, Jun. 2005.

10. Sunyog M.: Lean Management and Six-Sigma yield big gains in hospital's immediate response laboratory: Quality improvement techniques save more than $400,000. *Clin Leadersh Manag Rev* 18:255–258, Sep.–Oct. 2004.

11. Castle L., Franzblau-Isaac E., Paulsen J.: Using Six Sigma to reduce medication errors in a home-delivery pharmacy service. *Jt Comm J Qual Patient Saf* 31(6):319–324, 2005.

PART TWO

LEAN APPLICATIONS

CHAPTER SIX

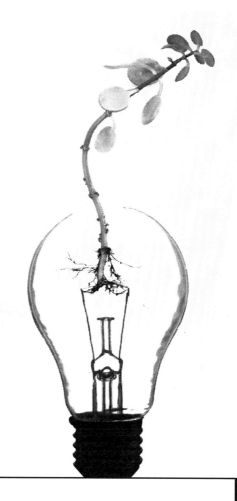

LEAN IN THE HOSPITAL SETTING:

USING VALUE STREAM MAPPING AND

KAIZEN TO IMPROVE PATIENT CARE AT

VIRGINIA MASON MEDICAL CENTER

Case At a Glance

THE ORGANIZATION:	Virginia Mason Medical Center, Seattle
THE LEAN PROJECT:	Hospital Unit Nursing Cells
THE TOOLS USED:	Value stream mapping, *kaizen*
THE OUTCOME:	Nurses have more value-added time with patients.

The Organization

Virginia Mason (VM) Medical Center, founded in 1920, is a nonprofit regional health care system that combines a primary and specialty care group practice of more than 400 physicians with a 336-bed acute care hospital in Seattle. VM operates seven regional clinics and manages Bailey-Boushay House, a nursing residence and Adult Day Health program for people living with HIV and AIDS. VM also has an internationally recognized research center, Benaroya Research Institute at Virginia Mason. The organization has about 5,000 employees, admits approximately 16,000 patients a year, and serves more than one million outpatients.

Lean Beginnings and Purposes

Virginia Mason is one of the first institutions in the United States to study and implement the use of Lean thinking in health care. In 2000, the board of directors challenged VM leadership to take a close look at services and the care environment. Through a strategic planning process, board members continued to ask if everything the organization did was centered on the patient. As a result of this introspective review, the team determined that processes and procedures were designed more for the benefit of physicians and staff. This finding prompted leaders to look for a management method that could help overhaul the system and help transform health care.

Through peers at Boeing, Virginia Mason discovered the Toyota Production System (TPS) and Lean methodology. In 2002, a team of 36, including the chairman of the board and the entire senior leadership team, traveled to Japan to witness firsthand the miracles of TPS/Lean. More than 190 VM employees have gone to Japan, including leadership, management, physicians, nurses, and other front-line staff; and all 5,000 employees have received some level of Lean training.

After the first trip to Japan, VM leadership returned to Seattle and adopted Lean not only as a pilot project or process improvement initiative, but rather as the organization's singular management methodology. The organization named this program the Virginia Mason Production System (VMPS). VMPS is now a foundational element of the VM Strategic Plan (shown in Figure 6-1) that appropriately places the patient at the top of the pyramid and is now part of the institutional culture at Virginia Mason. VMPS helps drive everything the organization does to improve quality, enhance patient safety, and provide exceptional care to the patient.

Number of Events and Projects Done

At Virginia Mason, Lean and VMPS are not projects or singular events. VMPS is the organization's management method that embraces a process of continual improvement and zero defects. It is used daily to identify and eliminate waste and inefficiency in processes that are

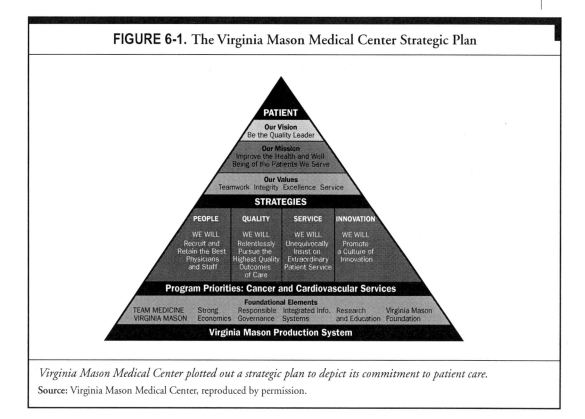

FIGURE 6-1. The Virginia Mason Medical Center Strategic Plan

Virginia Mason Medical Center plotted out a strategic plan to depict its commitment to patient care.
Source: Virginia Mason Medical Center, reproduced by permission.

part of health care. As of 2007, more than 4,000 Virginia Mason employees had participated in about 550 rapid process improvement (RPI) workshops, which are weeklong activities to identify waste or defects and improve on processes to create efficiency and value for patients.

The Lean Project:
Enhancing Time with Patients for Inpatient Hospital Nursing

Project Goals
In one of the many Lean-related efforts at Virginia Mason, a team of nursing leaders identified a series of challenges that were causing nurses to spend too much time doing tasks unrelated to direct patient care. The goals of the project were to reduce the amount of non-value-added work and increase the amount of value-added time between patients and nursing staff. The team accomplished this work through a series of efforts including *kaizen* events, value stream mapping, and RPI workshops.

Project Beginning and End Dates
The work in the nursing units began in November 2005 and has included ongoing efforts. The team looked at a series of ways to improve the value-added time nurses had with patients. Most of the major initiatives that were studied through an RPI workshop were implemented within VM's three-month time frame. Participants in an RPI workshop were typically from one unit or

floor. Those efforts that were particularly successful in enhancing value to the patient were then implemented on additional floors. Nursing unit enhancements continue today.

Baseline Data

Studies conducted by the Murphy Leadership Institute, Rand Corporation, and the Institute for Healthcare Improvement have all revealed that 35% of hospital activity failed to meet the definition of "value added." VM's own observations demonstrated that nurses spent only 32% of their time in direct patient care.

Team Members and Roles

An eight-member team consisting of nurses, patient care technicians (PCTs), and a patient participated in an RPI workshop to evaluate the work flow of nurses on the telemetry unit, the time spent walking, the time spent in patient rooms, and the overall value of time spent with the patient. All team members were responsible for identifying challenges and generating ideas that could help improve the value-added time nurses spend with patients.

Training on Processes and Tools

VM operates a department called the *Kaizen* Promotion Office (KPO). This team of employees possesses enhanced skills in Lean methodology and implementation. KPO team members work closely with staff throughout an RPI workshop and other *kaizen* events to provide on-the-job Lean training and use of tools.

The Tools Used

For this initiative, an RPI workshop was used to identify challenges in work flow and to determine solution strategies. (A related *kaizen* workshop idea form is shown in Figure 6-2.) Team members tracked the current state using stopwatches to measure time spent with patients and time spent on non–direct patient care. Walking steps were counted to determine how many miles nurses walked to do their jobs (including locating supplies, relaying information, updating electronic charts, and so forth). A value stream map was created to visually identify waste in the process and evaluate the flow of nursing care.

Project Steps

The primary methodology used was an RPI workshop. Although the central event in this process was the weeklong workshop, the process also involved eight weeks of preparation and check-in evaluation points at several-month intervals.

Old Process/Plan Versus New

In the current state (old) process:

- Patient assignments were made randomly based on patient acuity.
- There was no collocation of nursing assignments.
- Supplies and equipment were located a distance from patient rooms.
- There were no visual controls to signal the status of the nurse, PCT, or patient.

FIGURE 6-2. *Kaizen* Workshop Idea Form

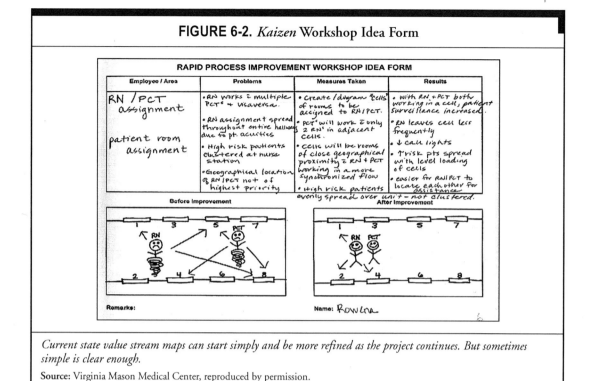

Current state value stream maps can start simply and be more refined as the project continues. But sometimes simple is clear enough.

Source: Virginia Mason Medical Center, reproduced by permission.

- Nurses and PCTs were engaged in patient care activities in parallel processes rather than in synchronization.
- There was a heightened risk for faulty or absent communications.
- Lack of routine patient surveillance resulted in missed changes in a patient's status.
- There was incidental overtime related to missed breaks and lunches.

In the new process:

- Staff members were organized into "cells" with responsibility for a group of patients in adjacent rooms. Patients in the same room were no longer split between nurse and PCT teams.
- Nursing assignments were now made based on geography. In the previous arrangement based on acuity, the disproportionate amount of time spent in wasted movement (searching for people, walking for supplies) more than offset the time it took to care for higher-acuity patients.
- Documentation and staff-to-staff reports on patient care were conducted inside patient rooms rather than outside rooms in the traditional way. Doing so allowed for more time with patients and families while records and reports were being completed. It also allowed for more patient-centered care, giving patients and families more time with caregivers to ask questions and offer and receive information.
- The formerly chaotic, frenzied telemetry unit became calm and controlled. Nurses now spend a greater part of their time at the bedside caring for their patients, resulting in significant changes in key quality indicators such as decreased falls and skin breakdowns, dramatic reductions in patient call-light usage, and increased patient and staff satisfaction.

Educating Staff on New Processes

In the beginning of implementation, staff members were hesitant to complete documentation in the patient room and needed follow-up reminders and encouragement to adhere to the new standard in their work. Daily check-in huddles with staff managers allowed staff to air concerns and receive support on difficult issues.

Biggest Obstacles

The execution of this standard work was a bit difficult at first. Initially, staff members were reluctant to change their familiar practices and embrace a new standard work flow that placed them back at the patient's bedside rather than at the nurses' station or utility rooms.

Concerns of feeling uncomfortable while performing work at the bedside have diminished over time, along with some feelings of exhaustion at shift ends. Nurses have felt more satisfied with their care delivery and nursing practice. Management's daily presence on the unit has been critical in moving this work forward, and 90-day follow-ups have demonstrated further gains and sustenance of new practices. The key to overcoming these obstacles was maintaining ongoing conversations with nurses and PCTs to listen to their concerns. Ultimately, the benefits to the patients, positive patient feedback, and diminished work stress helped win over many of the early objectors.

Easier than Expected

Some of the solutions came more easily than VM expected. In retrospect, team members see that sometimes the simple solutions are the best solutions, such as charting patient information in the patient room rather than at the nurses' station. But work culture and "the way it has always been done" mentality can often get in the way of progress.

The Outcome

Measurable Outcomes

- Staff members are completing their work during the morning cycle in two hours rather than four, a 50% reduction in time.
- The percentage of patients activating call lights during one monitored shift dropped from 5.5% to 0%.
- Lean improvements in the nursing units have increased the percentage of time nursing teams spend in direct patient care from 32% to 90%.
- The amount of distance that a nurse walks during a shift has been reduced by 85%.
- Nursing hours per patient day, which had been running over budget at an average of 9.0, have fallen below budget to 8.36.
- Because staff members are routinely able to get to their breaks and lunches, a 2% decrease in overtime has resulted.
- Increased time spent at the bedside allows for more frequent surveillance, which has led to multiple positive patient outcomes.
- The unit has had a decrease in patient falls, in skin breakdowns, and in pressure ulcers. This reduction in complications can potentially lead to reduced length of stay.
- Staff satisfaction improved with qualitative feedback from nurses and PCTs in several units.

FIGURE 6-3. Nursing Cell Results at 90 Days

"Nursing Cells" – Results > 90 days

- **RN time available for patient care = 90%!**

Before	After
• RN # of steps = 5,818	846
• PCT # of steps = 2,664	1256
• Time to the complete am cycle of work = 240'	126'
• Patients dissatisfaction = 21%	0%
• RN time spent in indirect care = 68%	10%
• PCT time spent in indirect care = 30%	16%
• Call light on from 7a-11a = 5.5%	0%
• Time spent gathering supplies = 20'	11'

This figure shows the results of Virginia Mason Medical Center's Lean project, the redesigning of nursing work units into "cells," 90 days after implementation. The project included work done by both registered nurse (RN) and patient care technician (PCT) personnel.

Source: Virginia Mason Medical Center, reproduced by permission.

- The patients' and nurses' experiences of the caring relationship and caring outcomes are being tracked subjectively by patient interviews.

Other 90-day results are shown in Figure 6-3 above.

Time to Achieve Outcomes

Within 90 days, results could be quantified. Within eight months of the initial RPI workshop that focused on the telemetry floor, work spread throughout all medical floors.

Time and Resources Saved

In addition to those outcomes noted earlier in the "Measurable Outcomes" section, nurses also saved nearly 5,000 walking steps a day—a significant savings of both time and personnel resources.

Wastes (*Muda*) Saved

Savings in motion, overprocessing, inventories, and waste of talent (staff time) resulted.

Communicating Progress to Staff

In addition to collecting data and reporting measurable outcomes to staff, VM produced a case study document titled "Getting Back to Nursing" that outlined the teams' multiple nursing-related Lean initiatives and the measurable outcomes. This document was distributed to staff and others within the organization to share the teams' successes.

Lessons Learned

There has been quite a bit of work devoted to this project, including implementation, endurance, training of staff, holding staff accountable to standard work, and going back to remeasure. The greatest lesson learned was the importance of maintaining diligence in sustaining results and not walking away too soon.

The staff's daily life changed greatly, ultimately leading to greatly increased satisfaction. If you can get staff over the hump of resistance to change, employees naturally transform their views regarding value and allow themselves to change and enjoy the results.

Last Words

VMPS/Lean work has fundamentally transformed the way nursing care is provided at Virginia Mason Medical Center. This work has enabled nurses to get back to the bedside and do the work they were called to do.

CHAPTER SEVEN

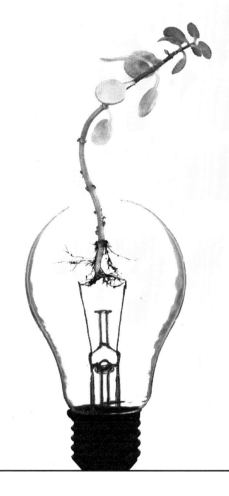

LEAN IN THE HOSPITAL SETTING:

USING SIX SIGMA TO IMPROVE
NURSE RETENTION AT MORTON
PLANT MEASE HEALTH CARE

Case At a Glance

THE ORGANIZATION:	Morton Plant Mease Health Care
THE LEAN PROJECT:	Six Sigma DMAIC Nursing Retention Project
THE TOOLS USED:	Six Sigma, value stream mapping, error proofing
THE OUTCOME:	Lower turnover rate, reduction in excess loss of registered nurses

The Organization

Morton Plant Mease (MPM) Health Care is a not-for-profit health care organization dedicated to providing community-owned health care services that set the standard for high-quality, compassionate care. MPM's vision is to be recognized as the preeminent community-focused health care organization in Florida. MPM employs more than 7,200 team members. Morton Plant Mease Health Care is comprised of the following hospitals: Morton Plant, Clearwater; Mease Dunedin, Dunedin; Mease Countryside, Safety Harbor; and Morton Plant North Bay, New Port Richey. MPM is part of BayCare Health System, a family of health care providers consisting of the nine leading not-for-profit hospitals in the Tampa Bay region.

Lean Beginnings and Purposes

Six Sigma DMAIC/Design for Six Sigma (DFSS), Lean, and Work-Out were all launched in this organization in March 2005. Prior to that time, MPM used the Plan, Do, Study, Act (PDSA) approach to performance improvement. Although the PDSA approach was effective in some cases, MPM found it lacking in that it was not consistently data driven, and when data were used, there was a reliance on "averages" rather than looking at variation in a process. Without valid, statistically relevant data, the organization felt it was difficult to know what to change in a process to achieve the most improvement, and it was equally difficult to sustain the gains. The Six Sigma DMAIC/DFSS, Lean, and Work-Out tools provided a solid frame-work for leading teams through process improvement. In addition to Lean and Six Sigma, described in Part 1 of this book, the following definitions may be helpful:

- *Six Sigma DFSS (Design for Six Sigma)* is a methodology that addresses building a new product, service, or process so that it is error free and meets customer requirements with Six Sigma precision. DFSS uses the DMADV approach (Define, Measure, Analyze, Design, Verify). Obtaining the "voice of the customer" is critical in this methodology as new and efficient processes are built.
- *Work-Out* is a methodology originally created by General Electric to empower front-line team members to improve organizational processes. It is a "fast" approach to process improvement that requires getting the right individuals in the room to "work out" a solution, while led by a trained facilitator. Data may not be required, but team tools such as affinity diagrams, fishbone diagrams, impact/effort matrices, and action plans are commonly used.

MPM's use of Lean, Six Sigma DFSS, and Work-Out created a common language throughout the organization for performance improvement. MPM phased in these methodologies, starting first with DMAIC, Lean, and Work-Out and then adding Six Sigma DFSS in 2007–2008.

The combined-methodology approach has evolved and is considered critical to project success. MPM incorporated Lean components into the DMAIC projects to get the best of both approaches. Currently, a DMAIC project may include such Lean tools as value-added versus non-value-added process mapping, 5S, elimination of *muda* (waste), and *poka-yoke* (error proofing), among others. Lean events also are staged separately if it is strictly a Lean project. Finally, in an appropriate situation, only a few of the Lean tools themselves may be used to apply to a given problem that doesn't need a "project." MPM has found this approach to be a useful way of integrating these methodologies into their culture.

Number of Events and Projects Done

Throughout all of BayCare, the following numbers of projects have been done:

Type of Project	No. of Projects
DFSS	8
DMAIC	172
Lean	30
Work-Out	194

The Lean Project

Project Goals

Goals were to improve timing, decrease defects, and reduce waste.

Background

The nursing shortage being experienced throughout the United States is a problem for all health care organizations. Recruitment strategies are important, but perhaps even more key to diminishing the impact of the nursing shortage is developing and retaining team members (employees) who are already a part of the organization. Some nursing turnover is normal in the marketplace, but excessive turnover, especially when it is a dramatic change from prior performance, merits an intense review. Although applying the DMAIC methodology might not normally be thought of as an approach to this type of problem, MPM needed to take action. This was a unique Six Sigma project (DMAIC) due to several reasons that included the following:

- The strong human resource component. This was not a manufacturing floor or a typical hospital process for DMAIC application.
- The project team itself could develop solutions. Due to the nature of the solutions, however, they could not implement them themselves. Yet by the Improve Phase, the team became a catalyst for change, making solution recommendations that were implemented by others.
- Because hospital volume is seasonal (with heavier work load in the winter months), there is normal variation in the data from month to month, and turnover also naturally varies.
- The need to keep the scope at a high level. Typically, a DMAIC project is focused on a single metric with a smaller scope. In this case, changes or improvements that were made impacted the lives of team members, and MPM recognized that if you do for one, you must do for all. Therefore, in most cases, the organization could not "pilot" changes on, for example, a single unit to ensure that the changes worked.

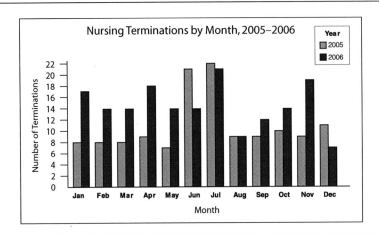

FIGURE 7-1. Nursing Terminations by Month, 2005–2006

The organization focused on reducing the 42-nurse annual overage, shown here, to a more reasonable level, targeted at 12.6.

Source: Morton Plant Mease Health Care, reproduced by permission.

Defect Definition
Registered Nurse (RN) termination from MPM (both voluntary and involuntary)

Problem Statement
In 2006, MPM lost 42 more RNs to termination compared with the previous year, resulting in an increase of the RN turnover rate from 10.9% in 2005 to 14% in 2006. Increased vacancies resulted in frustration to the nursing staff; additional costs related to paying agency, travelers, and pool RNs; and the potential to impact RNs' efficiencies in providing patient care on the unit.

Objective Statement
Decrease the additional number of RNs terminating from MPM by 70%, from 42 to 12.6 (per year), thereby reducing turnover rate and vacancy rate by November 2007.

Project Beginning and End Dates
The project began in September 2006. Project improvements were implemented over a period of time, with all implemented by June 2007. The project was kept "open" until November 2007 to monitor results via numbers of terminations per month and turnover rates. Some terminations are considered normal, and more nurses leave during certain months (June is a higher month because children get out of school), so MPM needed to keep the project open to ensure that gains sustained were indeed real.

Baseline Data
The baseline was calculated as the number of terminations for the year 2006 in excess of the number of terminations for 2005, or 42 (as shown in Figure 7-1 above). The corresponding

turnover rate baseline was 14% for 2006. The team members found it easier to focus "one nurse at a time" on the actual number of terminations as they were "getting their hands around" the project, making it easier to grasp the problem at hand.

At MPM, a 70% improvement is considered a "breakthrough improvement" in Six Sigma projects; therefore, a target of reducing the 42 additional nurses lost in 2006 (as compared to 2005) to 12.6 nurses lost in 2007 (as compared to 2005) was set.

A financial amount of $16,444 per termination was established as the financial savings for every nurse saved, based on time-to-fill-position data and the amount of agency/pool/contract expense the organization was incurring to backfill. This figure did not factor in other costs such as retraining new team members to replace those terminated; thus, it was a conservative estimate of the total financial savings.

Team Members and Roles

Project Champion: Lisa Johnson, M.S., M.S.N., R.N.-C.N.A.A., Chief Nursing Officer, Vice-President Patient Services

Project Leader: Pam Guler, M.H.A., Administrative Fellow, Six Sigma Master Black Belt

Process Owner: Sharon Collotta, R.N., Director, Team Resources

Team Members:

Bedside Nursing	Nurse Managers/Directors	Other Areas
Sue Roscoe, R.N.	Sharon Fietz, R.N.	Carol Dimura, R.N., Education
Jamie Wendel, R.N.	Joan Clow, R.N.	Bev Witkowski, R.N., Recruiting
Kimberly Loucks, R.N.	Marcia Albanese, R.N.	Mary Quinn, Systems Analysis
Victoria Bosi, R.N.	Shannon Hancock, R.N.	Linda Menken, Finance
Karen Brown, R.N.	Jackie Munro, A.R.N.P.	
Nicole Ford, R.N.		

This project was championed by Lisa Johnson, chief nursing officer (CNO). She was involved throughout the project, providing guidance and support, removing barriers, and recognizing the efforts of the team. Many of the improvement solutions involved her at a detailed level because she provided education back to the nurse managers as discoveries were made and the need for refresher education on such topics as fair scheduling of shifts and holiday/weekend hours was identified. MPM's president and CEO, Phil Beauchamp, was also extremely supportive in the improvement implementation. Under his name, the team implemented a major culture change systemwide by creating a "Meeting-Free Zone" where all employees are prohibited from scheduling meetings between 8:00 and 10:00 four mornings per week. This policy allowed managers to be on the units rounding more often on both their team members and patients. The bedside nurses identified this need, as they perceived that their managers were away in meetings too often. Data analysis provided evidence that nurse managers typically spent seven hours a week in standing meetings, not including additional ad hoc meetings scheduled as needed. MPM recognized that it was "a meeting-happy organization," and this practice changed from senior management down with the implementation of the Meeting-Free Zone.

Training on Processes and Tools

The team members received just-in-time (JIT) training on the Six Sigma DMAIC methodology provided from a structured curriculum led by Pam Guler, the Master Black Belt team leader. An overview of the methodology and what was to come was done at the beginning of the project; separate education for each phase of DMAIC was done as each phase was reached. The team reinforced learning by working with the hands-on tools during the given phase. Team members earned their Yellow Belts by doing all the training, participating in the project, and completing an online exam at project end. The Yellow Belt is a special recognition awarded by senior management to all eligible team members on an ongoing basis as projects are completed.

The Tools Used

Project Steps

The steps of the project are listed by phases as follows.

Define Phase
1. Establish the team with the right team members (bedside nurses were key).
2. Identify the process of a "terminating nurse" at a high level from nurse recruitment to termination via a tool involving Suppliers, Inputs, Process, Outputs, and Customers to determine Requirements (SIPOC-R). This identification allows everyone on the team to understand the process at hand in general terms and gets everyone on the same page.
3. Determine the defect definition, problem statement, and objective statement (see previously described project background and goals).

Measure Phase
1. Develop a detailed process flow diagram at each step of the process.
2. Based on the process, develop a data collection plan to do the following:
   - Obtain data on terminations (numbers listed by position, turnover rate, and vacancy rate) and reasons for termination from internal systems for review.
   - Obtain data on time in position prior to termination for each nursing job level: Clinical Nurse Resident (newly graduated nurses), and Clinical Nurse I, II, and III (how long nurses were staying in their position prior to termination and whether this could offer any insight into needs).
   - Obtain all available data on nurses' satisfaction with their jobs and supervisors through the use of the following surveys:
      * Team Member as Customer Survey results (a yearly survey that that MPM conducts)
      * Nursing New Measures Survey
3. Begin to narrow down potential reasons for termination and key steps in the process that may impact retention via the fishbone (*Ishikawa*) diagram and cause-and-effect matrix.
   - Obtain more data as necessary based on team tools (for example, the team suggested that fairness in scheduling holiday/weekend coverage might be a reason for frustration and leaving the organization, so a survey was distributed to nurses to determine if that was indeed a concern).

Analyze Phase

1. Using graphical analysis, Pareto charts, and review of survey results, analyze the data previously obtained.
2. Continue to narrow down the key areas for focus via the data analysis from Step 1 and additional team tools, such as Failure Modes and Effects Analysis (FEMA, described in further detail in the subsequent section "Old Process/Plan Versus New").
3. Identify "critical *X*s": the top items that have the greatest impact on nursing retention, based on all prior steps and tools. These items become the focus for the Improve Phase. The critical *X*s in this case were the manager/nurse relationship, respect toward the nurse, work load/stress, child care, and education for new nurses.

Improve Phase

Based on the critical *X*s identified and the original fishbone listing many of these *X*s, the team brought in more bedside nurses to brainstorm improvements specific to these items. In effect, this action used the PDSA tool. The team did the following:

1. Categorized and prioritized brainstormed solutions
2. Reviewed solutions with the CNO and nursing directors, who were all very supportive
3. Developed a 24-step action plan for implementation. Assistance was sought from outside the team due to the nature of many solutions.
4. Implemented the solutions
5. Communicated to the nurses themselves what the team had done so that the nurses would be in the loop. This inclusion was important for this project.
6. Monitored and evaluated the monthly termination data to determine improvement over time. This phase was not closed until the team was confident of this improvement.

Control Phase

In this phase, the team did the following:

1. Established an ongoing monitoring mechanism for the data along with regular reports to management on progress
2. Established a control plan for action if the turnover metric begins to decline. That calls for immediate communication to all nurse managers as well as analysis.
3. Closed the project

Realization Phase

For 12 months after the project closes, the team reports the monthly turnover data to the senior management team. If it is out of line with the target, an action plan is required, but that has not happened so far in this realization period. Along with the turnover rate, the team is monitoring a 12-month rolling average of terminations per month to level for natural variation in monthly terminations.

Tools Used

- SIPOC-R diagram: A team tool to get everyone familiar with the process at a high level (Figure 7-2 on page 105).

- Process flow diagram: Detailed from recruitment to termination.
- Data collection plan: Based on the process flow diagram, a listing of all data that would potentially be helpful in the project, and how to collect them (Figures 7-3, 7-4, 7-5, 7-6 on pages 105–107).
- Failure Modes and Effects Analysis (FMEA, Figure 7-7 on page 107): Based on the fishbone diagram and the top scoring steps (only) in the cause-and-effect matrix, analyze each key step and determine how that step might fail with regard to retaining the nurse, how severe that failure would be, how often it occurs, and how easy it is to detect. A numerical score is assigned to each item, and an overall risk priority number is calculated for each step.
- Fishbone (*Ishikawa*) diagram: A cause-and-effect tool used by the team to anecdotally brainstorm why a nurse might leave.
- Cause-and-effect matrix: A team tool spreadsheet that has the team take every step in the detailed process flow and assign a numerical score (on a scale of 1 to 10) to how important that step is in retaining the nurse.
- Minitab data analysis: The sample provided in Figure 7-8 on page 108 is a graphical analysis of the time-in-position data for the clinical nurse residents (new nurses). Of those who terminated, it was found that they were leaving at five months, which led to further brainstorming on why the organization was not keeping these new nurses after their orientation period. The team determined that the preceptor program needed major change, and a new initiative was launched to create orientation units for new nurses with lower nurse/patient ratios and "super preceptor" nurse educators overseeing all activity. New nurses now learn in these units that are conducive to consistent teaching prior to going to their home unit. This process provides an experience with no variation in content and with experienced nurse mentors.

Old Process/Plan Versus New

The critical causes of nurse termination that the tools identified included the nurse manager-to-RN relationship, equity in the unit for shift/holiday/vacation assignment, proper precepting of new nurses in an environment that is conducive to learning, stress and hectic pace on the unit, relationships with patients and de-escalating stressful situations, and respectful relationships with physicians. Innovative solutions included the following:

- An increased focus on the relationship between the nurse manager and the nurse manager's employees via a program of progressive education in retention techniques for nurse managers. The nursing directors were trained to be "retention coaches," or experts in this area.
- A Meeting-Free Zone was implemented every day, for two hours each morning, for all employees throughout all four hospitals. This policy allowed the nurse manager to be in the unit and not away in meetings. This was a culture change that has proven highly successful. Senior management also make rounds during this time to be in touch with employees.
- Policies on fair scheduling were reinforced throughout the hospitals.
- Two orientation units for new nurses have been established, so that they may be precepted by expert RNs who enjoy teaching while they work.
- Focus groups were created where nurses may relieve stress, and a daily program allowing nurses to break away for a massage, soft music, and relaxation techniques during lunch was implemented.

FIGURE 7-2. SIPOC-R (The Process)

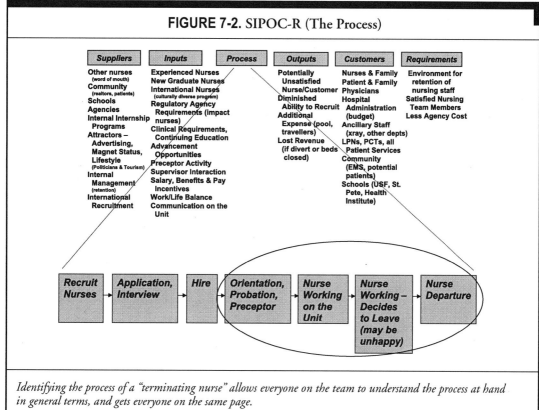

Identifying the process of a "terminating nurse" allows everyone on the team to understand the process at hand in general terms, and gets everyone on the same page.

Source: Morton Plant Mease, reproduced by permission.

FIGURE 7-3. Internal Staffing Survey Results

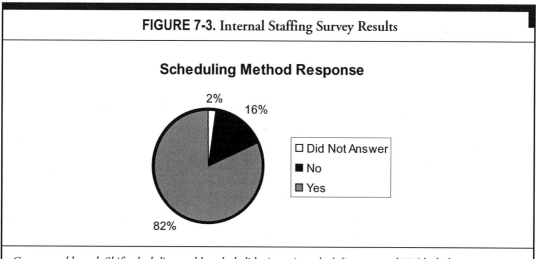

Concerns addressed: Shift scheduling and length, holiday/vacation scheduling surveyed 704 bedside nurses to gather true need for change in these areas. The Take-Away: Some units prefer self-scheduling and others prefer master scheduling. The data results were not strong enough to make a broad change at this time.

Source: Morton Plant Mease, reproduced by permission.

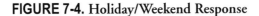

FIGURE 7-4. Holiday/Weekend Response

Holiday/Weekend Response

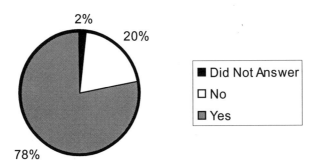

- ■ Did Not Answer
- □ No
- ■ Yes

The Take-Away: Reeducate nurse managers on the importance of fair handling for holiday and weekend scheduling. Ensure that the policy is followed.

Source: Morton Plant Mease, reproduced by permission.

FIGURE 7-5. Shift Length Response

Shift Length Response

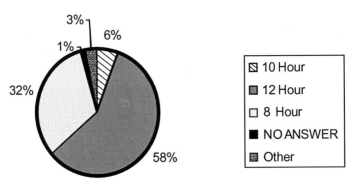

- ▧ 10 Hour
- ■ 12 Hour
- □ 8 Hour
- ■ NO ANSWER
- ▦ Other

The Take-Away: Continue to offer a mix of shift lengths throughout the system, rather than going to "all 12s" as other hospitals have done.

Source: Morton Plant Mease, reproduced by permission.

FIGURE 7-6. Summary for Length of Stay (LOS) in Last Nursing Position

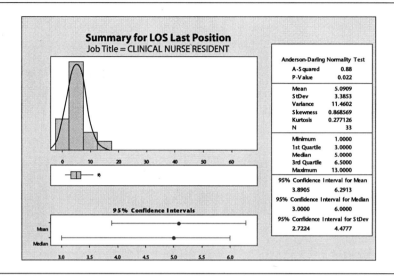

Analysis of length of tenure for newer nurses who terminate. Mean number of months of employment = 5.09 months. The Take-Away: Action was needed to ensure that new nurses received consistent precepting and were prepared for their nursing role.

Source: Morton Plant Mease, reproduced by permission.

FIGURE 7-7. Failure Modes Effects Analysis (FMEA)

Process Function	Potential Failure Mode	Potential Effects of Failure	S E V	Potential Cause(s)/ Mechanism(s) of Failure	O C C	Current Process Controls	D E T	R P N	Recommended Action(s).
The highest value process steps from the C&E matrix.	In what ways might the process potentially fail to meet the process requirements TO SATISFY/ RETAIN NURSES?	What is the effect of each failure mode on the NURSE?	How severe is the effect to THE NURSE?	How can the failure occur? Describe in terms of something that can be corrected or controlled. Be specific. Try identify the causes that directly impacts the failure mode, i.e., root causes.	How often does the cause or failure mode occur?	What are the existing controls and procedures (inspection and test) that either prevent failure mode from occurring or detect the failure should it occur? Should include an SOP number.	How well can you detect cause of FM?	SEV x OCC x DET	What are the actions for reducing the occurrance, or improving detection or for identifying the root cause if it is unknown? Should have actions only on high RPN's or easy fixes.
Nurse interacts with his/her manager on the unit.	Manager may not be available due to meetings, some more visible on all shifts, new nurses may not feel welcomed, bond or relationship with existing staff may not be strong.	Unhappy Nurse, Lack of feeling support on the unit.	9	Too many meetings, Manager may be occupied with other duties, Manager may not check in on off shifts, Manager may not be aware of nurse Team Members concerns or have a strong relationship, Trust may not be built, Manager may not know what the nurse needs/wants		Expectations of managers are discussed at leadership meetings, More assistance is needed	9	729	Implement a Meeting Free Zone 4 mornings a week so that mornings can be freed up to round on units and interact with Team Members & Patients, Ask the Nurses what they want their manager to do via survey, Implement TalentKeepers on-boarding and retention techniques for building trust.

Only the highest-risk priority number steps provide the critical Xs to address in the Improve Phase.

Source: Morton Plant Mease, reproduced by permission.

FIGURE 7-8. Control Phase Results

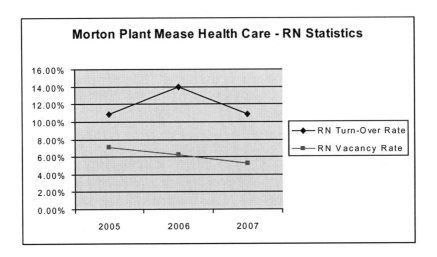

The team monitored January through December 2007 to determine improvement (and observe seasonality). The excess loss of registered nurses (RNs) was reduced to 12 over 2005 (target 12.6). The graph shows the RN turnover rate and vacancy rate. The annualized conservative savings totaled $609,000.

Source: Morton Plant Mease, reproduced by permission.

- Techniques for de-escalating stressful situations with patients were communicated via education.
- A social event was held in a fun setting for 400 nurses and physicians and featured a physician band led by the president of the medical staff. This event was used to enhance working relationships between physicians and nurses.
- Communication of these changes to the bedside RN was key via nursing newsletters.

Educating Staff on New Processes

Most of the education was provided to nurse managers. The CNO conducted this education in her "all manager" meetings and worked with her directors and managers for follow-up.

Biggest Obstacles

Getting the needed time for project team meetings with MPM's extremely busy bedside nurses, nurse managers, and nursing directors was a challenge. To address it, the team had to be flexible and creative with meeting times and willing to change to meet immediate staffing needs. This time issue prolonged the project by about a month, but the accommodation was necessary because having the bedside nurses at the bedside delivering care is paramount.

There was initial concern that the Meeting-Free Zone would add a few more hours onto the day because it might necessitate meeting later. To prevent that from happening, communications focused heavily on the topic of streamlining meetings and included good meeting tips:

making meetings meaningful, inviting only those who are necessary, converting face-to-face meetings to conference calls, and having and sticking to agendas. The overall culture has changed; people feel comfortable questioning whether they need to be at certain meetings. The response to this program has been overwhelmingly positive. The nurse managers are now able to be and stay in their units each morning, making rounds with their team members.

This was not your "typical DMAIC." It was a major issue and a "world hunger" project. Creatively adapting the tools of the methodology to meet the needs for this topic was challenging but worked out well.

Easier than Expected

Sometimes in a major DMAIC project, it is difficult to get buy-in and commitment to the solutions for improvement that the team develops. With this project, where so much needed to be done outside the small team, MPM anticipated that it might have challenges but did not encounter any. The team became a catalyst for change through the "top down" leadership of the CNO and her nursing directors. Their leadership and the key participation of bedside nurses throughout the project were a recipe for success.

The Outcome

Measurable Outcomes

MPM reduced the excess loss of RNs to 12 (with a target of 12.6.) Turnover rate dropped from 14% in 2006 to 10.9% in 2007, one of the lowest in the area.

Time to Achieve Outcomes

The project took 8 months to implement plus an additional 5 months to monitor success, a total of 13 months.

Time and Resources Saved

Although this factor was not quantified, MPM achieved a significant time savings in recruitment, orientation of replacements, and disruption of the unit.

Wastes (*Muda*) Saved

Wastes saved included the following:

- Waste of intellect and rework. Losing good people means that you have lost their knowledge and skill set, and then you need to start over and train the next team member.
- Waste of processing. The process is now more streamlined to eliminate excessive meeting time.

Cost Savings

Very conservative annual savings from this change, based strictly on reduction in agency/contract backfill costs, was set at $609,000 ($16,444 per termination). If the cost of training new nurses is considered in the equation, a truer estimate would be over $1 million per year.

Communicate Progress to Staff

Mass communication e-mails and articles in internal newsletters announced the Meeting-Free Zone. To promote these changes and ensure that everyone was aware of them, all changes were communicated via the internal nursing newsletters.

Lessons Learned

This project required flexible scheduling in how the organization approached Six Sigma project meetings to meet the needs of the nursing staff. This approach has rolled over into other subsequent projects, making it easier for team members to get involved in projects because meeting agendas and timing are handled more efficiently. For example, MPM has moved to a "rapid DMAIC" meeting approach that requires fewer meetings spread out over several months, but each meeting lasts a longer period of time (for example, two to three hours). The organization found that it was easier for a nurse to get away for a planned span of time than to leave for a one-hour meeting once per week, which had been the previous practice.

CHAPTER EIGHT

LEAN IN THE AMBULATORY CARE SETTING:

USING *KAIZEN*/2P TO REDESIGN
AN OUTPATIENT PAVILION AT
ADVOCATE GOOD SHEPHERD

Case At a Glance

THE ORGANIZATION:	Advocate Good Shepherd Hospital, Barrington, Illinois
THE LEAN PROJECT:	Redesign of the Outpatient Pavilion
THE TOOLS USED:	Rapid improvement events (RIEs, or *kaizen*), 6S training (Good Shepherd employs a sixth S: safety), waste walks, *gemba* (front-line) walk, value stream mapping
THE OUTCOME:	Improved patient and provider satisfaction through improved flow and increased customer value; improved use of clinical space square footage, patient walking distance, supply flow, and patient privacy

The Organization

Advocate Health Care, based in Oak Brook, Illinois, is the largest fully integrated not-for-profit health care delivery system in metropolitan Chicago. Advocate has eight hospitals with 3,500 beds and a home health care company among its more than 200 sites of care. More than 24,500 people work at Advocate, with more than 4,600 affiliated physicians. Advocate's primary academic and teaching affiliation is with the University of Illinois at Chicago Health Sciences Center.

Advocate Good Shepherd Hospital, one facility in the Advocate system, is a 183-bed acute care hospital in Barrington, Illinois. The hospital has more than 600 physicians on staff representing 35 medical specialties, and it handles approximately 12,000 admissions, 30,000 emergency department (ED) visits, and more than 115,000 outpatient encounters per year. Although it is a community hospital, Good Shepherd offers high-tech and tertiary services such as open heart surgery, image-guided radiation therapy, and robotic surgery.

Lean Beginnings and Purposes

Advocate Health Care began its Lean journey in January 2007. Simpler Healthcare℠ helped leadership identify value streams on which to focus and provided high-level training across the Advocate system in February 2007. In March 2007, Advocate Good Shepherd completed an enterprise value stream analysis (EVSA) to identify and prioritize value streams for improvement, subsequently selecting the ED and the revenue cycle.

Later that spring, the team performed a value stream analysis (VSA), taking three days to evaluate each of the two identified value streams. They identified multiple Just Do Its to immediately effect change in areas for which solutions were easy to implement and made common sense. They also identified a list of projects to undertake that would take up to 30 days to complete. A number of rapid improvement events (RIEs) were also determined, the first of which was held in June 2007.

Number of Events and Projects Done

Sixteen RIEs had taken place by early 2008, including five RIEs focused on facility design. These events included redesigning the hospital's on-campus imaging facility, labor and delivery unit, and outpatient pavilion, and undertaking a new bed expansion project.

The Lean Project

Project Goals

Advocate Good Shepherd's objective was to redesign its high-volume, 21,000-square-foot Outpatient Pavilion housing 10 different clinical services and various administrative functions. The primary reason for action was to improve patient, physician, and associate (employee) satisfaction through improved flow and increased customer value. The final design also improved clinical space square footage, patient walking distance, supply flow, and patient privacy.

Project Beginning and End Dates

The project reflected in this case study fell within the value stream of the revenue cycle and pertained to the *physical layout* of the outpatient pavilion. In its current state, the team quickly realized that the physical layout was a barrier to customer value and flow. The current state had a negative impact on patient satisfaction, associate satisfaction, patient flow and efficiency, and the ability to increase patient capacity.

Baseline Data

Space in the current outpatient areas was not properly configured. Problems included the following:
- Areas with high volume and low treatment times farthest from entrance
- Room size and layout not appropriate based on procedure type
- Treatment areas cluttered and unorganized
- No room for expansion
- Poor seating arrangements in two waiting areas and a main lobby
- Significant portion of the pavilion occupied by nonclinical space

Patient satisfaction, as measured by a Press Ganey score, received the lowest rating for privacy, and patients became confused and lost in the process of trying to find the clinics. Employees were also dissatisfied, citing supply location and the consistency of inventory as poor. They also thought spaces allocated for their personal belongings, staff lunches, and breaks were minimal (as shown in Figure 8-1 on the next page).

Team Members and Roles

A weeklong RIE using the Lean 2P (Process Preparation) technique was led by Simpler Healthcare (http://www.simplerhealthcare.com). Ten Good Shepherd associates representing a variety of disciplines attended the full-week RIE. Another 8 to 10 ad hoc team members attended for a few hours or a half day during that week. In all RIEs, Good Shepherd strives to include a "one third, one third, one third" mix: one third front-line people, one third people trained in Lean, and one third "outside eyes." The broad array of roles represented included strategic planning, facilities (architect and construction), nursing (a front-line gastrointestinal nurse, a front-line pain nurse, the nurse manager of the outpatient pavilion), administration (the director of ambulatory services for Good Shepherd), and consulting.

Good Shepherd put into place a policy whereby senior management would be expected to participate in at least two RIEs a year. The goal was not merely to complete a number of improvement projects but rather to transform the culture; thus, it is recognized that the more

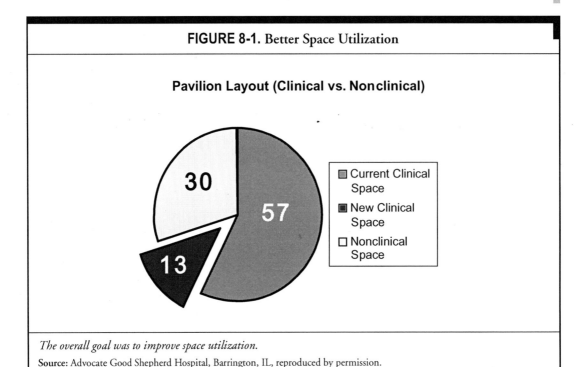

FIGURE 8-1. Better Space Utilization

Pavilion Layout (Clinical vs. Nonclinical)

- Current Clinical Space
- New Clinical Space
- Nonclinical Space

30

57

13

The overall goal was to improve space utilization.

Source: Advocate Good Shepherd Hospital, Barrington, IL, reproduced by permission.

associates are involved, the greater the cultural transformation will be over time as associates bring experiences and learnings back to their work areas.

Because the RIEs involved front-line people, time was tracked and appropriately budgeted within the Division of Operations Improvement's cost center. Administration set the tone of the RIE. Everyone was poised to implement changes as quickly as possible even during the week itself. Administration also provided support in the chance of a barrier or roadblock. Continental breakfast, lunch, and snacks were provided each day to boost energy and morale. At the end of each day, daily management updates were given by the team leader to administrators, who then were given the opportunity to comment on the team's progress and to ask questions. At the end of the week, the completed A3 (described later in this chapter) was presented to the administration and all Good Shepherd associates in a "report-out" forum. Each team member was required to participate in the report-out and was given a portion of the A3 to describe during the presentation. The week-ending report-out serves two functions. First, it gives the team members a chance to display the results of their weeklong session to their colleagues, and second, due to administration's presence, it gives the results weight that is carried into subsequent stages of production.

Training on Processes and Tools

Simpler Healthcare *senseis* (master teachers) educated the team in the Lean 2P process design and facilitated the week's programming. The definition for 2P used in this project was "Process Preparation" as opposed to 3P (Production Process Preparation) because the existing space was going to be used but redesigned. Prior to the weeklong RIE, preparatory meetings were also held.

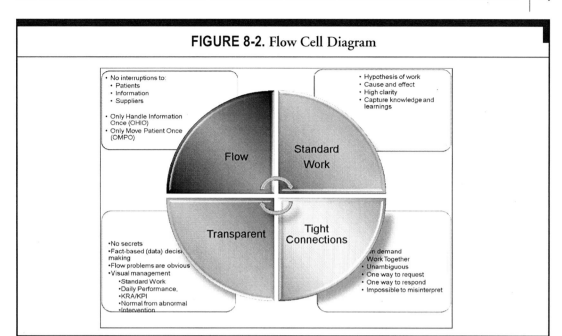

FIGURE 8-2. Flow Cell Diagram

This diagram portrays both the project objectives identified and the tools selected for major use accordingly.

Source: Advocate Good Shepherd Hospital, Barrington, IL, reproduced by permission.

The Tools Used

Project Steps

The project objectives and tools used are represented as shown above in Figure 8-2. Steps taken and measures used to create quantitative data for each clinical area included *takt* time (discussed in the next section), average number of patients seen per day, square feet, revenue per square foot, associate walking distance, patient walking distance, patient satisfaction scores, and projected growth for services over the next five years.

In addition, a waste walk performed in each area identified 55 areas for improvement. Frontline staff received 6S training (sort and scrap, straighten, scrub/sweep, standardize, sustain, safety) to garner immediate gains in efficiency prior to construction.

Takt time

Takt time (TT) was defined as the heartbeat or pace of the system or process. A relationship of available time for tasks or services to be performed was plotted against the number of tasks or services to be accomplished (demand). The TT was calculated using the following formula:

If available time = 8 hours, or 480 minutes, less planned breaks and lunch (approx 45 minutes), then available time = 435 minutes.

Demand = 50 patients per day. TT = 435/50 → 8.7 minutes → thus, every 8.7 minutes a patient should be served.

TABLE 8-1. Utilizing Lean Process Preparation (2P) to Redesign an Outpatient Pavilion

Use of a 7-3-1 layout design matrix provided the following benefits:
- Key criteria ranked, giving the most important criteria the highest weight
- Seven layouts (for the whole pavilion) created focus on key criteria
- Scored using scoring methodology: weight x score (1, 3, or 5, with 5 being best)
- Three optimal layouts identified
- One "hybrid" layout created that encompassed the best aspects of the other layouts
- Hybrid layout scored against key criteria

Key Criteria	Weight	Design 1	Design 2	Design 3	Design 4	Design 5	Design 6	Design 7
Privacy/Confidentiality	4	12	20	12	12	12	12	12
Patient Flow	7	7	21	21	21	21	21	7
Minimize Walking	5	15	15	15	25	15	5	25
Accessibility	6	30	6	30	18	6	18	30
Openness to Environment	1	3	1	3	3	3	3	3
Cost to Implement	2	6	2	2	10	2	2	2
Flexibility/ROI	3	9	3	3	15	15	15	9
Total Score		**82**	**68**	**86**	**104**	**74**	**76**	**88**

Source: Advocate Good Shepherd Hospital, Barrington, IL, reproduced by permission.

Using the quantitative data matrix as a guide, the group used the 7-3-1 layout alternative process to create seven different alternatives to the current state. This process is reflected in Table 8-1 above and Figure 8-2 on the previous page.

Tools Used

Seven key criteria (privacy, patient flow, minimization of walking, accessibility, openness to environment, cost, and flexibility/return on investment) were weighted to judge seven alternative layouts. The group voted to give patient flow the most weight and assign openness to environment the least weight. Each of the seven options was then scored on each of the seven key criteria using a layout decision matrix. Stakeholders from areas not represented on the team were consulted in a WIIFM ("what's in it for me") session to review ideas and offer suggestions for improvements. A final "hybrid" design was developed using the best elements of each of the redesigns. A construction table was created and phased over three distinct time periods. Figure 8-3 on the next page shows the A3 template.

Using the A3 Template
A3 heading and boxes 1, 2, and 3 are completed prior to the seven-week event cycle

Quality Check—Questions to Be Answered
- Does event align with future state value stream?

FIGURE 8-3. A3 Template

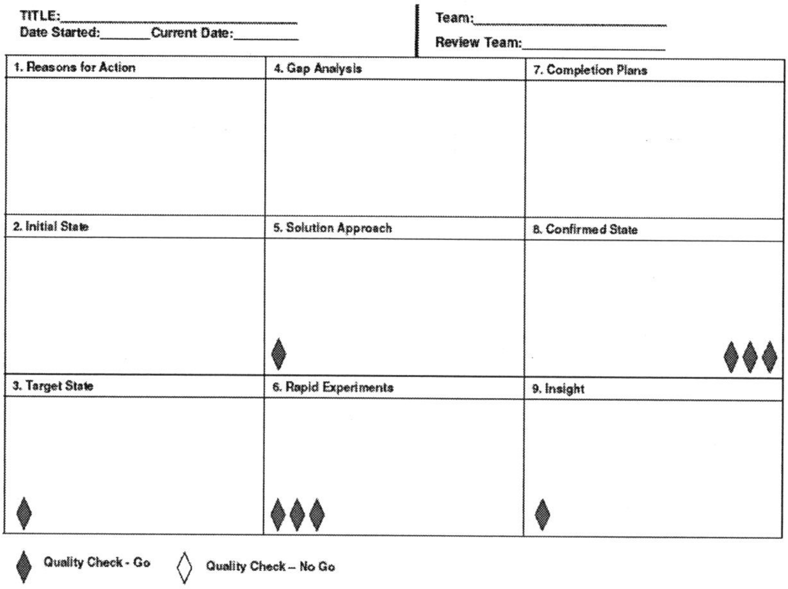

The A3 tool helps in problem solving by facilitating the organization of thoughts and priorities.

Source: Simpler Healthcare^{SM}, reproduced by permission.

- Are true north metrics defined and are targets achievable? Have appropriate team members been identified?

Boxes 4 and 5 are completed during three-week prep cycle

Quality Check (Beginning of Third Week of Preparation)
- Has gap analysis been completed with well-thought-out cause chains?
- Does solution approach in box 5 link well with the gaps identified in box 4?
- Has three-week prep cycle been completed with demonstrated results?
- Are team members committed to the team for the entire week?

Box 6 is completed during days 1, 2, and 3 of the event week

In-Process Quality Checks (Daily Management Updates)
- Are rapid experiments achieving desired results/learnings?
- Is team following the solution approach identified in box 5?
- Can roadblocks be removed?
- Is daily checklist being completed?

Boxes 7, 8, and 9 are completed during day 4 of the event week

Out-Brief Quality Checks
- Is completion plan in place with names, dates, and outcomes defined?

- Are confirmed state metrics in place and do they validate the target state?
- Does confirmed state show what good looks like?
- Do insights show key learnings and identify future opportunities?

30-, 60-, 90-Day Quality Checks
- Is completion plan on track/complete?
- Are true north metrics moving toward the target state?
- Has improvement been sustained?
- At 90-day point, does review team agree to close out A3?

Other tools used
- 6S training (workplace organization)
- Waste walks (identification of waste in current operations)
- *Gemba* (front-line) walk (focus on process indicators and gain insight from users)

Old Process/Plan Versus New

In the old process of facility design, front-line staff members were not involved. With the process used, many voices were heard, so the end product was both associate and patient focused. Not only were front-line staff members represented on teams, but two weeks prior to the RIE, surveys were posted to ask front-line staff members and physicians for their comments, concerns, and suggestions. The actual design was finished faster because the design was well thought out ahead of time, reducing change requests and subsequent costs.

Biggest Obstacles
- Getting buy-in: Allowing scheduling to accommodate team members so that they could attend the full-week RIE was possible only because upper management bought in to the project.
- Practicalities: Having someone from construction at the RIE gave context to comments and costs. Occasionally some participants wanted to jump to spend dollars on technology or additional staffing to solve problems. The *senseis* stressed "creativity before capital."

Easier than Expected

Simpler Healthcare introduced the teams to "try-storming," which involves experimentation in solution design. Having data made things easier. Data points for all clinical areas were posted for comparison—for instance, growth in cardiology compared with growth in pain management. Data also made it easier to compare travel distance for patients, and that helped drive decisions. Whereas the data collection itself was first seen as an obstacle, having that data became a boon to the project.

The Outcome

Measurable Outcomes
- Patient walking distance for the highest-volume clinical service area (lab services) decreased 66.3%, or more than 300 feet walked per patient; similar improvements were made in other areas as well (see Figure 8-4).

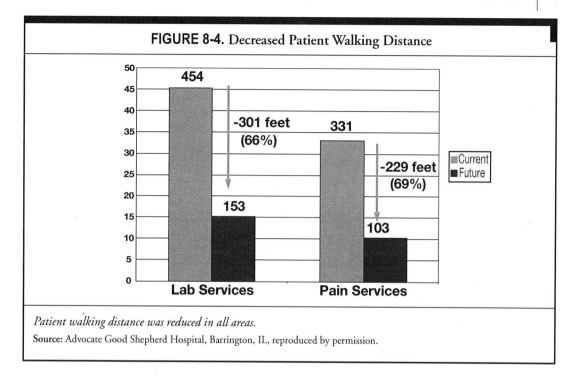

FIGURE 8-4. Decreased Patient Walking Distance

Patient walking distance was reduced in all areas.

Source: Advocate Good Shepherd Hospital, Barrington, IL, reproduced by permission.

- One waiting room was eliminated and converted to clinical space.
- Clinical space was increased from 57% of the pavilion square footage to 70%, an addition of 2,500 square feet of clinical space (see Figure 8-5 on the next page).
- This new space increased clinical capacity and brought in additional revenue per year, resulting in a payback period of less than three years to recover construction costs.
- A centralized storage area was created, thus reducing staff walking distance to find supplies common to multiple areas.

Time to Achieve Outcomes

The new outpatient pavilion design was completed in 2007, and the first two phases were expected to be completed in 2008.

Time and Resources Saved

Walking distance was the biggest waste saved. Better space utilization meant greater allocation of clinical space. New storage space arrangements saved staff motion.

Communicating Progress to Staff

A number of communication outlets were used to keep staff abreast of proposed changes coming out of RIEs and the progress of implementing changes following RIEs. These outlets included 30-, 60-, and 90-day team leader updates at report-outs and in official newsletters. Posters in the evolving pavilion also served this purpose. Everyone was excited about the project and was aware of stages of completion. Administration was kept abreast through daily management updates, and staff members not involved in the RIE were invited to attend the report-out at the end of the RIE week.

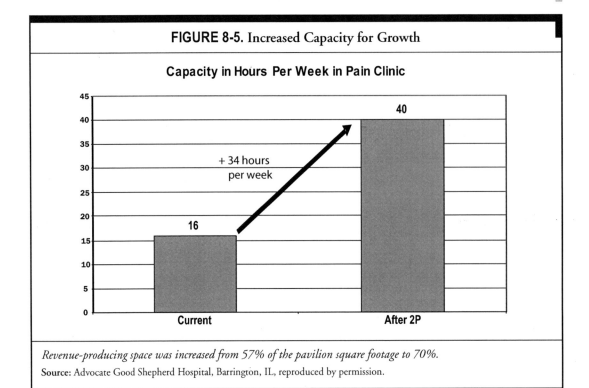

FIGURE 8-5. Increased Capacity for Growth

Capacity in Hours Per Week in Pain Clinic

Revenue-producing space was increased from 57% of the pavilion square footage to 70%.

Source: Advocate Good Shepherd Hospital, Barrington, IL, reproduced by permission.

Lessons Learned

- Using the 2P process can significantly help in the brownfield design of clinical space. (*Brownfield* is space that is currently occupied and will be redesigned, such as the outpatient pavilion project described in this case study; *Greenfield* design is currently unoccupied space requiring a design from the start, such as the creation of a new bed tower.)
- Data-driven decision making made final decisions much easier.
- All future space planning should utilize the 2P process before ground is broken.
- This process supports the notion of "creativity before capital," which helped keep costs down and forced the group to think more creatively before jumping to solutions involving additional staffing or technology.
- Major decisions should utilize a Rank "x" Weight matrix considering all criteria.
- Coming and going during an RIE can greatly distract the process and is actively discouraged.
- Establishing other "game rules" before an RIE is important: show up on time, no side conversations, and so forth.
- Front-line staff members designed the work area they use and thus achieved greater utility.

CHAPTER NINE

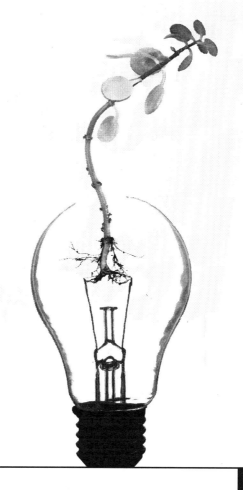

LEAN IN THE AMBULATORY CARE SETTING:

USING *KAIZEN* TO IMPROVE

PRIMARY CARE PATIENT FLOW AT

DENVER COMMUNITY HEALTH SERVICES

Case At a Glance

THE ORGANIZATION:	Denver Community Health Services, Denver Health and Hospital Authority, Denver
THE LEAN PROJECT:	Primary Care Care Flow
THE TOOLS USED:	Value stream mapping, 5S, rapid improvement events (*kaizen*), Just Do Its (JDIs), and 2P and 3P events
THE OUTCOME:	Improved patient flow, increased patients' value-added time while in the clinic, increased provider and staff productivity and access to care

The Organization

Denver Community Health Services (DCHS) is the primary care component of the Denver Health and Hospital Authority located in Denver. DCHS provides primary care services and urgent care services to low-income individuals in Denver through a network of 8 community health centers, 12 school-based health centers, 2 urgent care centers, and a hospital-based women's care clinic. Services provided are comprehensive primary care services, dental services, mental health consultation, and urgent care services. The number of unduplicated patients served in 2007 was 103,826 individuals in 371,129 visits. The department has about 650 full-time equivalent (FTE) employees, including 65 physicians, 46 allied health practitioner FTEs, 16.8 dental FTEs, and 80 registered nurses.

Lean Beginnings and Purposes

Lean management began at Denver Health in 2006. It was adopted as a framework for improving processes at an organizational level, and all departments have participated in the activity.

Number of Events and Projects Done

- DCHS has conducted more than a dozen rapid improvement events (RIEs) as part of its Lean activity.
- It has conducted three value stream mappings, which provided a timetable for its RIEs.
- DCHS has conducted a number of 5S activities to improve site organization.
- It has conducted three 3P (Place, People, and Processes) or 2P (People and Processes) activities. (Note: Consultants and teams sometimes use different names for 2P and 3P.) DCHS is opening a new clinic, so with 3P it designed space for that clinic, the processes that will be used, and the people it will employ. In the 2P events, because the space already exists, DCHS will design for only the people and processes.
- In addition, DCHS has conducted a number of other projects as part of its initiative.

The Lean Project

Project Goals

The primary goals of the Lean initiative in DCHS were to improve patient flow, thereby reducing patients' cycle time through the clinic, to increase the value-added time that patients spend in the clinic, and to increase the productivity of providers and clinical staff, thereby increasing access to care and users.

Project Beginning and End Dates

The project began in January 2006 and did not have a scheduled end date. By 2008, DCHS was in its third year of Lean activity using the same primary goals.

Baseline Data

The essentials of baseline data for the project are the following:
- Patient visits per provider session = 8.2 visits per session
- Adjusted panel size per FTE = 1,400 patients per FTE
- Patient cycle time exceeded 60 minutes

Team Members and Roles

There were various members on the teams carrying out RIEs, value stream mapping events, or 2P/3P events. In general, each team was composed of approximately eight individuals with a team leader usually selected from outside the clinic of focus. Team leaders were individuals who had received training in Lean process improvement activities, and teams were made up of a team leader, a facilitator with extensive training in Lean activity and the support of a *sensei*, one or two process owners for the event who were responsible for carrying out the activities outlined at the end of the event, and approximately six other team members, half of whom came from the clinic of focus and half of whom attended as outside participants either because of their expertise or because of their ability to think creatively about process improvement.

Managers and clinical directors in community health at DCHS often were trained as "Black Belts." Black Belts are individuals who have been given extra training in Lean methodology and Lean tools, and they frequently served as one of the team leaders for an RIE or 2P/3P event. Management also provided encouragement, coaching, and problem solving during an event week and heightened awareness of the events to emphasize the importance of Lean as a way to improve processes within the organization.

Training on Processes and Tools

For each RIE, the team members received training on Lean processes and tools at the beginning of the event week, usually the morning of the event week. Ideally, one third of the team members will have been involved in an RIE previously.

The Tools Used

Project Steps

The primary steps in the Lean project were first to conduct a value stream mapping to identify waste within the system, a reason for action, and measurable outcomes from the value stream events. Value stream mappings are used as a way to outline the RIE, 2P events, or other projects to be done over the next six months to a year. Figure 9-1 on the next page shows a preliminary

---

FIGURE 9-1. Preliminary Current State Value Stream Map

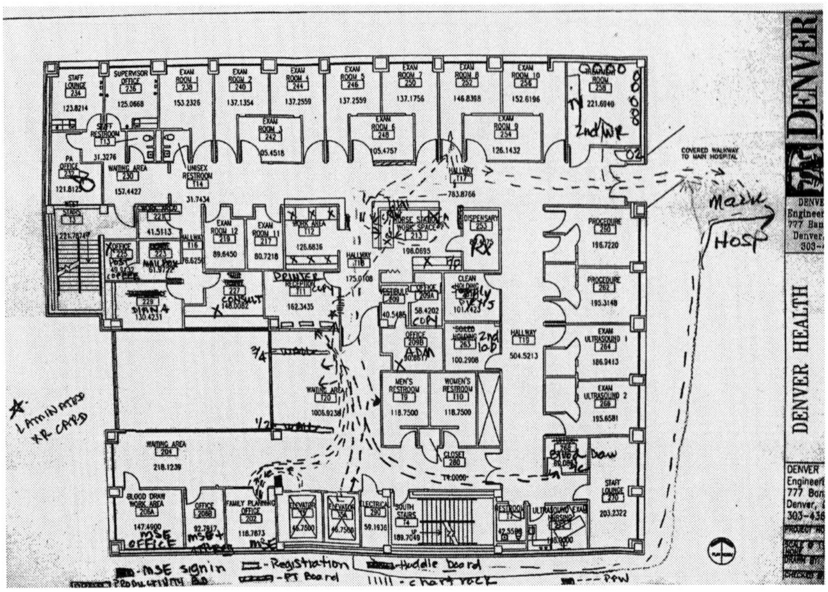

Value stream mapping is used to identify waste. This hand-drawn map shows the current state value stream.

Source: Denver Community Health Services, Denver, reproduced by permission.

---

hand-drawn current state value stream map of clinic flow. Figure 9-2 on page 125 shows a spaghetti diagram, a tool used to illustrate the movement of people in creating a current state map. A formalized current state map is shown in Figure 9-3 on page 126.

Tools Used

In addition to value stream mapping, the specific Lean tools used were 5S organizing events, RIEs, JDIs, Lean projects, 2P events, and 3P events.

Old Process/Plan Versus New

A number of changes in processes were implemented as a result of the Lean activities over the previous two years. Significant changes were to do the following:

- Reduce appointment types to a single type of appointment versus a mixture of complex appointment types.
- Implement "forced calling" of telephone calls to allow for a significant reduction in calls being lost.
- Reroute new appointment calls to a central call number to distribute new appointments throughout the system.
- Change the sequential processing of patients to a more parallel process involving the medical office assistant and the provider to eliminate the duplication of service and improve clinic processing time.

FIGURE 9-2. Spaghetti Diagram

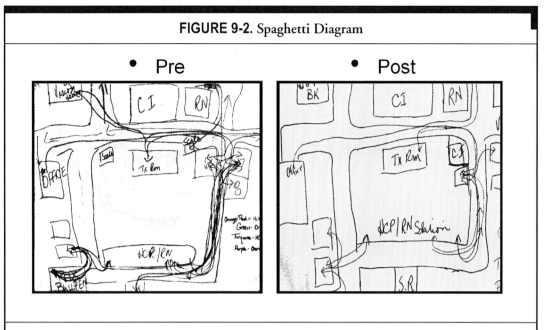

• Pre   • Post

This tool shows the movement of people. The "Pre" figure highlights wasted motion and indicates where improvement could be made. The "Post" figure shows a more streamlined process. Spaghetti diagrams are used to create value stream maps.

Source: Denver Community Health Services, Denver, reproduced by permission.

- Reduce the number of missed appointments by developing and implementing a systemwide late or no-show policy.
- Perform a by-site measurement analysis of individuals' activity to reallocate tasks to the most appropriate team member.

Educating Staff on New Processes
Staff education was conducted both through monthly presentation of successes and lessons learned to departmental leadership and through dissemination of that information from leadership to the clinic-based teams.

Biggest Obstacles
The biggest obstacles for the project were the organization's inexperience with using Lean as a tool for process improvement and its initial attempt to carry out activities at multiple clinic sites in a year. In the second year of Lean activity, DCHS decided to focus its activities on only three clinics and repeatedly conduct events at those clinics to improve Lean successes.

Easier than Expected
It was easier than expected to enlist staff in accepting the process because DCHS already stresses reliance upon staff members to develop their own interventions. Staff buy-in has led to a great interest in conducting Lean activities at various sites.

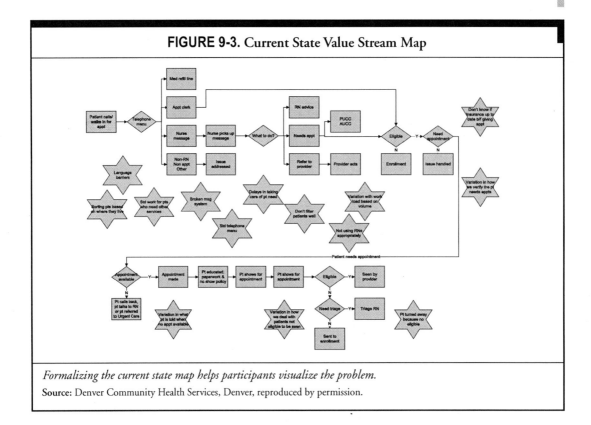

FIGURE 9-3. Current State Value Stream Map

Formalizing the current state map helps participants visualize the problem.

Source: Denver Community Health Services, Denver, reproduced by permission.

The Outcome

Measurable Outcomes

The most measurable outcome of the activity pertained to individual and team productivity. The average visit per provider increased from 8.2 to 9.8 visits per session. In addition, the average panel size per FTE increased to more than 1,500 per provider.

Time to Achieve Outcomes

It took approximately two years to achieve this outcome, although there has been a gradual improvement over that period of time.

Time and Resources Saved

Because of the Lean activity, resources were not saved; however, efficiency improved by increasing production and thus increasing numbers of patients served.

Wastes (*Muda*) Saved

The waste that was saved has been primarily in the area of unused talent. By looking at individuals' activity within the clinic flow process, the team can now more appropriately assign tasks to individuals within their job classification, and management has essentially increased the complexity of work for support staff members who were not working to their full potential.

Cost Savings

The total revenue generation through this project was in the range of $1.8 million over a two-year period. This is a "hard dollar" amount associated with increased visits, increased charges, and increased revenue.

Communicating Progress to Staff

The success of the initiative is being communicated to staff through leadership meeting presentations, news announcements, and ongoing discussion at both provider and management staff meetings.

Lessons Learned

For the next time DCHS conducts a Lean event, it learned to do the following:

- Increase the level of communication specific to the Lean events it is doing.
- Focus activities on a limited number of clinics instead of multiple clinics simultaneously.
- Identify and use a systematic process for rolling out processes to new sites without having to conduct Lean activities at those sites.
- Better sustain their gains. With 5S activities, this is always a concern. DCHS learned that the most successful sites are those having a manager who keeps on top of things and can caution employees by saying, "We're slipping!" as a reminder to continually engage in maintaining order and organization.

Last Words

It is important for people who will conduct Lean activity to understand that there is a need to educate the leaders of the initiative on Lean methodology. DCHS uses a consultant firm to assist with its activities, to train facilitators, to train leaders (Black Belts), and to consult and guide these facilitators and leaders in their Lean journey. In addition, Lean as a methodology emphasizes process improvement at the staff level and therefore empowers staff members to make changes. Ensuring that changes are made quickly and that staff members feel that they are empowered to do so without interference from managers is an important concept. It may require change in management style and beliefs within an organization.

Advanced Lean Thinking

CHAPTER TEN

LEAN IN THE HOME HEALTH CARE SETTING:

USING *KAIZEN* TO IMPROVE

HOME MEDICAL EQUIPMENT

PICKUP AT NORTHEAST

HEALTH

Case At a Glance

THE ORGANIZATION:	Northeast Health, Albany, New York
THE LEAN PROJECT:	Northeast Home Medical Equipment Pickup Process
THE TOOLS USED:	*Kaizen*, process mapping, standard work
THE OUTCOME:	Better communication among departments; standardized work flow within each department; balanced work load for drivers, leading to fewer missed deliveries and fewer customer complaints

The Organization

Northeast Health (NEH) is a regional, comprehensive, not-for-profit network of health care, supportive housing, and community services. NEH was formed in 1995 by the merger of Samaritan Hospital and The Eddy network and was joined by Albany Memorial Hospital in 1997. In 2007, Sunnyview Rehabilitation Hospital of Schenectady became an affiliate of The Eddy. NEH provides an array of health care services, including primary care, acute care, surgical care, imaging, laboratory, emergency services, behavioral health, chronic and long term care, rehabilitation, community-based care, adult day care, home care services, assisted living, and retirement housing. NEH serves 22 counties in the capital region of upstate New York, cares for approximately 175,000 people annually, and employs 5,000 people.

Lean Beginnings and Purposes

In 2003, the NEH board of directors and quality committee performed a yearlong study on process improvement methodologies and eventually recommended Lean based on its simplicity and ability to involve the front line. One affiliate of NEH chose a *kaizen* event to address equipment pickup. Although customer satisfaction was high per survey results, complaints continued to be received associated with the pickup of equipment not being completed as promised (as scheduled). Other issues included the number of redundant calls responding to the patient's/family's question, "Why hasn't the pickup occurred?" and the amount of driver overtime and general driver dissatisfaction (the need for a balanced work load).

Number of Events and Projects Done

As of July 2008, the organization had conducted 36 five-day *kaizen* events and 7 mini *kaizen* events. It also had conducted 12 learning labs (intensive, full-focused, one-year projects with one department at a time) and started 2 more in 2008 that were still in progress.

The Lean Project

Project Goals

The project goals were to do the following:

- Meet customer (patient, family, referral source) expectations and promises
- Increase driver satisfaction through balanced work load

- Reduce all driver overtime
- Reduce redundant calls coming in from customers (intake calls)

Project Beginning and End Dates

This particular *kaizen* event was held March 7–11, 2005. Data collection and role/participation designation began two months prior to the *kaizen* workshop. Implementation continued through April 2005.

Baseline Data

When customers had equipment such as beds and commodes that were no longer needed (for instance, the patient had died), company contacts (NEH does a lot of work with hospice) were making promises to families for pickup at certain times. Promises were then changed by customer service and again by drivers. Promises were made, changed, and changed again, and as a result, customers became frustrated and distressed. Drivers' schedules were changed by others, or they changed them themselves. When families had to hurry the process, they would overly disassemble equipment (unscrewing components that did not need to be taken apart each time) and put it aside, creating more reassembly work for drivers down the line. The baseline was ultimately calculated in terms of the number of promises that were late, which was 80 out of 215, or about 37% of the time (63% on time).

Team Members and Roles

NEH has found that a *kaizen* team of 22–25 people is not as productive as a smaller team; but in a team of only 3 or 4 people, the perspective is limited. Over time, they found that an ideal number of participants is 10–12. The participation for this *kaizen* event included a facilitator, management, a driver, a customer service intake representative, a warehouse worker, and the head of distribution. The team leader was Irene Magee, Vice President and Director of Northeast Home Medical Equipment. The facilitator was Tricia Brown, Vice President, Corporate Affairs, Northeast Health.

Having the director on the team and present during the entire week was key to the process. She was integral to pulling the process through and to eliciting information from participants, who were placed in a role that they were not accustomed to having. The director also described the big picture to help participants better refine the process. For the future, although it was difficult to pull drivers off their regular work load, NEH resolved to find a way to include more front-line staff, which would better facilitate conversation and insights among the team.

Training on Processes and Tools

NEH provided training in "chunks" or "snippets" woven throughout the *kaizen* week. Participants received training in Lean thinking for a couple hours on Monday morning, and on Monday afternoon and Tuesday morning, they were trained on waste elimination.

The first process map was drafted pre-*kaizen*—the high-level "30,000-foot" map was drawn as a table that blocks out process/movement. At the next stage, during the *kaizen*, the participants mapped out the details of their jobs associated with this process (second map), and then mapped the customer promise (the product) in the third map. Participants were then given a series of questions to "waste walk" the maps on paper.

Participants were also given training in ideal state and Toyota Production System (TPS)/Lean tools on Wednesday of the *kaizen* workshop week.

The Tools Used

Project Steps

As illustrated in Figure 10-1, the steps included the following:
1. Identify the opportunity.
2. Form a team and scope the project.
3. Analyze current state.
4. Define desired outcomes.
5. Identify root causes and propose solutions.
6. Plan and test proposed solutions.
7. Refine and implement solutions.
8. Measure progress and hold gains.
9. Share results.

Tools Used

Process mapping was used a great deal, and observation was integral to using the tool pre-*kaizen* and during the event. When NEH participants first began the *kaizen* event, they considered the product to be an item of home equipment: the bed. They initially mapped where that bed was, where it went, and where it waited. However, as work progressed, it became clearer that the product was the promise that had been made to the customer. Standard work was a tool they found especially useful to make pickup consistent.

Old Process/Plan Versus New

The group used a "parking lot" posted on a board throughout the *kaizen* so participants could keep "solution" ideas for later work. At first they discouraged thinking about solutions and focused on the current state, but by Wednesday afternoon, they were in the flow of thinking out of the box. Where some manufacturing firms are able to complete changes on Thursday of the *kaizen*, Northeast gave it 90 days. It projected 90% completion of the work plan within three months. The team called that an "after action review": The team went through the work plan and shared information on what worked and what did not.

Other implemented changes included the drivers' schedules. In the old process, they were given the whole day's schedule. Their changes to that, based on actual work flow and calls they made and received, meant that those schedules were constantly being changed. In the new process, drivers were given a four-hour "miniroute" and then they called out for the rest of the day's schedule. This process helped alleviate a lot of rework.

Another change was to move from using big box trucks to more fuel-efficient vehicles. Staggering start and end times allowed for better flow of loading.

Educating Staff on New Processes

To educate staff on new processes, NEH leaders held team meetings and full staff meetings

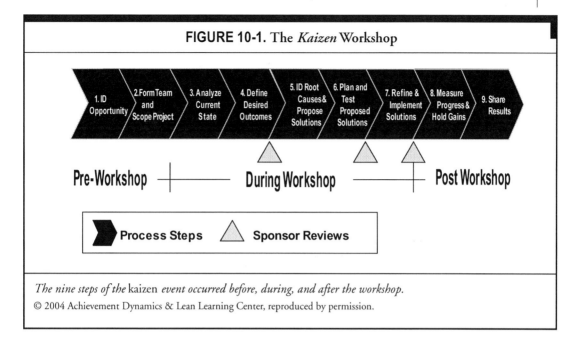

FIGURE 10-1. The *Kaizen* Workshop

The nine steps of the kaizen *event occurred before, during, and after the workshop.*
© 2004 Achievement Dynamics & Lean Learning Center, reproduced by permission.

and posted updates. They also held meetings with hospice regional directors to review project results and end outcomes.

Biggest Obstacles

After implementation, staff began to revert to poor communication and to distribution that allowed them to take both A.M. and P.M. routes in the morning. Also, it was difficult to take drivers off their scheduled routes to attend the *kaizen* event; pulling even five people amounted to a 25% reduction of the workforce.

Easier than Expected

- The ability to schedule an A.M. route and allow for fluid changes to P.M. routes improved work load issues.
- Better defined expectations of how and when staff communicates improved accountability.
- The *kaizen* event improved understanding and culture around the importance of promises.

The Outcome

Measurable Outcomes

The benefits after implementing the Lean solutions included the following:
- Balanced work load/expectations for drivers due to miniroutes
- Frequent check-ins by drivers
- Fewer missed promises
- Fewer carryovers
- Less frustration for families

- Frequent unloading of dirty equipment
- More cross-training of drivers
- Distribution now more "in the know"
- More standardized flow and standardization within each department
- Education for hospice
- Better communication
- Better understanding of one another's jobs
- For intake: defined process to describe time expectations
- For intake: standardization of order times/expectations for delivery and pickup

Time to Achieve Outcomes

Overall, this was a very successful *kaizen* event with the data sustained over the yearlong period. As shown in Table 10-1, the data revealed the following achievements:

TABLE 10-1. Pre- and Post-*Kaizen* Data		
	Pre-*kaizen*	**Post-*kaizen***
On-time pickup	63% measured pre-*kaizen* in first quarter 2005	84% for 9 months in 2005, sustained at 92% in 2006
Driver productivity	1.16 average deliveries per hour in 2004	1.21 average deliveries per hour in 2005
Miles driven per delivery	13 in 2004	12 in 2005
Costs	Dollar amounts not measured; however, in 2005, costs increased slightly due to rising gas prices and boxcar repairs	
Source: Northeast Health, Albany, New York.		

Time and Resources Saved

- 20% reduction time for delayed pickups that were pushed to the next day's schedule
- Savings in rework on intake-call tracking and follow-up when families asked, "Where is my driver?"

Waste (*Muda*) Saved

- Rework for intake and distribution schedules; data regarding specific savings were beginning to be collected in 2008.
- Miles were reduced even with the addition of vehicles returning to the warehouse midday, from 13 miles per delivery in 2004 to 12 in 2005.

TABLE 10-2. Lessons Learned
Choose the best players for the team and challenge them.
Maintain accountability.
Empower workers.
Consider and plan small vs. big projects for best success.
Select projects according to appropriate duration of time, clarity of product, availability or need for information, specific ownership, division stability, majority buy-in, capacity for implementation, and capacity for team and worker enthusiasm.

Cost Savings

NEH did not designate a dollar-amount measure beyond increased customer satisfaction that equaled added value, reduced miles traveled, and improved productivity.

Communicating Progress to Staff

Progress was posted on the Lean scoreboard and in newsletters and professional advisory reports. Progress was regularly discussed at team and full-staff meetings.

Lessons Learned

Table 10-2, above, summarizes the lessons learned from many events.

General Lessons

- Always challenge the team to consider what the product is instead of assuming what it is.
- *Kaizen* events that are sparked by a learning lab can have wonderful results. The need is known, and the players may have existing experience with Lean due to their roles with the lab.
- Choosing the "best" players for the *kaizen* event is not always in the best interest of the event. Sometimes the "difficult" staff can provide better input at the event and may become supporters by the end of the week.
- Accountability, in varying degrees, is a key component of each event.
- Keep to the *kaizen* steps through the week (even if a strong player changes direction and starts to jump to solutions). The steps work, and value exists in maintaining the order of the week.
- Multifacility events are "big" and do not lend themselves to the *kaizen* process (you don't get as much out of the maps as you would with a single facility). The product map can be difficult to create for multiple facilities. However, by the end of the week, these multifacility projects can be wonderful from many perspectives: sharing of best practices, getting to know peers and building respect, and capacitating rollout as a consistent process across a division.
- Many times, strong personalities are involved. It would be helpful, when possible, to learn about any issues prior to the event.
- The event can empower front-line workers to question how things are done. It is possible to improve upon work that has frustrated them to date.

- For learning labs, it often takes a few sessions to get the front-line workers to open up. That's okay; let the silence hang in the air when asking questions. It may get uncomfortable, but it allows them to open up.

Lessons Regarding Size of Project/Crossing Divisions

- Projects that cross multiple facilities and departments may be too large because it is difficult to focus the group on a single process (that is, there may not be enough time to focus on processes that are handled in different ways for each facility).
- Project solutions that affect more than one department (that is, nursing units) may be difficult to implement due to the sheer numbers of units and employees involved, especially if individuals from each unit were not involved in the *kaizen* event.
- Narrowing the project during the planning session is key, either through the number of steps involved or through narrowing the criteria to one type of diagnosis, form, or department affected. This is not to say that the solution can't positively affect other areas (for instance, other types of surgery), but the *kaizen* workshop is a short week and needs to stay focused within a manageable set of boundaries.
- Projects that can have a significant impact in a short period of time (that is, within 1–3 months) are geared for *kaizen* events.
- Projects that have a clear "product" are tailor-made for *kaizen* events; that is, you can visually see a patient or referral, for instance, move through the process, especially if it is "time" geared.
- However, NEH has also found tremendous benefit in the "world hunger" projects, those large projects that cross departments and processes and have "big picture" goals. Staff at various affiliates and divisions get to know one another and what they all do, they build a better understanding and respect for another component of the system, and these projects improve patient transfers between departments and affiliates across the system.

Lessons on When to Wait to Conduct a Particular Project

- If there is a sense that the *kaizen* project solutions may require long-term information systems (IS) support or other key support, it may be best to hold off if these resources were not originally budgeted. Some *kaizen* events have turned into large IS projects; that was not really the goal, but that is what tends to arise from focusing on the ideal state.
- If there is not a clear owner who will ultimately be responsible for the implementation of solutions, this should be a red flag at the planning session indicating that it may not be a good project to choose. The projects that have a clear owner, someone who can "run" with the project after the *kaizen* event, are more successful than those that do not.
- Choosing a *kaizen* approach in an area that expects to go through a major change, such as electronic or construction-related, is probably not appropriate because it may not be worth the week to dive into the current state when the current state is going to change shortly.

Lessons on When to Go After a Particular Project

Choosing a *kaizen* project in an area that has already done one is helpful from many aspects:

- Staff will be familiar with the structure, expectation, and goals.
- Staff can build upon a previous success (most processes within a department are linked and can expand the process improvements).
- *Kaizen* participants continue to send the Lean message and education to staff.
- *Kaizen* is a viable option when there is up-front agreement by the staff involved that there can be a better approach to an issue. This attitude can go a long way toward a successful *kaizen* event and implementation of the recommended changes. Getting buy-in is a natural step during the week, but if staff members truly do not believe there is a problem, there may not be time in the *kaizen* workshop to convince them, and they won't be supportive of the proposed changes.

Other Lessons Learned

The better the data going into the event, the more easily the process flows. If national benchmarks are available, that is a plus, but it is not a necessity. Sometimes staff members feel that if they are meeting a national benchmark, they are "golden," rather than viewing the ability to take further waste out of the process.

The after-action reviews are definitely needed. Some teams will work well if they have a strong process owner; but if not, they may flounder. Some teams focus on the key goal/measure but not on all the action plan steps. Other teams focus on the action plan and not the key goal. There is benefit to focusing on both the measure and the work plan. Teams need to be questioned about why they didn't pursue certain actions when it made sense at *kaizen* time.

Beware of solutions that require accountability by staff when team members are leery of the ability to implement. Accountability may continue to "bite" you in each event.

Kaizen is not the same as Lean. Continue stressing to teams that *kaizen* is only one tool of Lean and that Lean is a way of thinking.

Last Words

Even where a goal/measure was not met, all teams noted specific benefits related to the *kaizen* event (including, for instance, cleaning up their marketing, improving turnaround times, responding more effectively to the customer, knowing what other staff members do, and even better understanding their own jobs).

Teams were still excited about the *kaizen* six months or more after the event. They saw the value of the event and agreed that it was worth the resources.

Kaizen means something to staff now: It is a quick way to improve a process, a way to trigger a big project that has been on the plate for a while. Although it also means a substantial amount of work during the event and afterward, it is viewed as worth the effort.

CHAPTER ELEVEN

LEAN IN THE BEHAVIORAL HEALTH CARE SETTING:

USING A3 TO IMPROVE
CLIENT OUTCOMES AT
PYRAMID HEALTHCARE

Case At a Glance

THE ORGANIZATION:	Pyramid Healthcare, Inc., Pittsburgh
THE LEAN PROJECT:	Pyramid Performance Monitoring Program
THE TOOLS USED:	Toyota Production System (TPS) model, A3 problem solving tools
THE OUTCOME:	Shortened length of client intake process, discharge status, and length of stay; improved aggregate client outcomes

The Organization

Pyramid Healthcare, Inc., is a substance use disorder treatment system headquartered in Altoona, Pennsylvania, with programs located in western Pennsylvania (in or near Pittsburgh) and throughout Pennsylvania. Staff include a program director, four adult counselors, and one adolescent counselor. Pyramid worked with three programs in Pittsburgh that provide the services described in the following sections.

Adult outpatient: Pittsburgh, Birmingham Towers. Drug and alcohol assessment and treatment recommendation, dual diagnosis assessments and treatment, individual therapy, group therapy, family therapy, intensive outpatient programs (daytime and evening), partial hospitalization, professional consultations, educational presentations and workshops, and urine screening for drugs of abuse.

Adolescent care: Pittsburgh, Birmingham Towers. Group counseling, individual counseling as necessary, family component, focus on awareness and education on drug and alcohol issues, complete communications with referral sources, Youth At Risk program, educational presentations and workshops, and urine screening for drugs of abuse.

Men's 3/4 Way House. Staffing included six full-time staff members: one program director, one technical supervisor, and four full-time technicians. No treatment is provided at this facility, but the house provides recovery housing, as a starting-off point to begin a sober and lawful existence, to men who are involved in the Allegheny County Criminal Justice System. The house staff members help residents continue their recovery plans by participating in treatment, 12-step meetings, and recreational activities. This program, which is funded by a Pennsylvania Council on Crime and Delinquency (PCCD) grant, is a collaborative effort between Pyramid Healthcare, the Allegheny Department of Health, the Allegheny County Prison, and the Allegheny County Department of Human Services.

Lean Beginnings and Purposes

Planning for the implementation of the project began in early 2004. Data collection began in November 2004. The project was initiated from interest expressed by a stakeholder group that was convened by a local 501c(3) corporation, the Institute for Research, Education and Training in Addictions (IRETA), and from the experience of University of Pittsburgh School of Pharmacy faculty (coleader of stakeholder group) using Toyota Production System (TPS) applications in health care. Pyramid's CEO expressed interest in being a pilot site for the application of a substance use disorder treatment–based performance improvement approach using TPS principles.

Number of Events and Projects Done

This has been the only Lean project put into practice at Pyramid thus far, although the School of Pharmacy is engaged in applying other Lean applications within pharmacy practice and other areas.

The Lean Project

Project Goals

The goals of the project were to do the following:

- Provide real-time reporting to clinical staff regarding client status at intake and at 30-day intervals while in treatment to facilitate treatment planning and determine service needs for clients.
- Develop a reporting system that will identify areas in need of improvement/change to enable continuous program improvement. For example, reporting on the length of the intake process from initial client contact with the program to actual intake identified a time lag that could result in the loss of client entry into the program due to the excessive wait time.
- Illustrate the real time and aggregate progress of clients by reporting the change over time for four domains:
  1. Alcohol and/or drug use
  2. Criminal justice involvement
  3. Employment
  4. Housing

Aggregate comparisons were performed monthly between intake and discharge, comparing only those clients for whom both intake and discharge data were available. This parameter eliminated including the intake information from clients who did not finish the program and complete the discharge questionnaire, for example, those who were reincarcerated or who left therapy against medical advice.

Project Beginning and End Dates

- Adult outpatient data were collected from November 2004 through July 2007.
- Adolescent outpatient data were collected from February 2006 through July 2007.
- 3/4 House data were collected from August 2005 through July 2007, but program staff client-file reviews dated from January 2004.

Baseline Data

The team sought an instrument that could demonstrate to both private and public payers a reduction in multiple wastes. Pyramid standardized the intake assessment and embedded certain domains and measures that made clinical sense and would be required of the program by stakeholders. The baseline data included the following:

- Client baseline data collected at intake on the four previously mentioned domains. Demographic information was also collected at this time.
- Program baseline data obtained from the sites. These data included items such as length of stay in each level of care and an explanation of the intake process and the average number of days this process took.

Team Members and Roles

The team included the following members:

- Jonathan Wolfe, Chief Executive Officer of Pyramid Healthcare
- Mark Schmidhofer, M.D., Director, Cardiology Consult Service Cardiovascular Institute, University of Pittsburgh Medical Center (UPMC) Presbyterian
- Jody Bechtold, Director of Performance Improvement/Project Lead
- Stephanie Murtaugh, Clinical Consultant
- Jan Pringle, Ph.D., Director, Program Evaluation and Research Unit, School of Pharmacy, University of Pittsburgh

Management was extremely supportive of the project and its goals. The directors of the sites were instructed to fully participate in the implementation of the project at their sites. Administrators at the sites cooperated by doing the following:

- Approving staff time for meeting attendance and administration of the project
- Authorizing time during group counseling sessions for the completion of the periodic self-report data forms by the clients
- Making staff available for the clerical functions involved in data transfer and report distribution
- Sharing improvements in program operations as evidenced in the monthly reports to encourage staff acceptance of the processes advocated by the project

Training on Processes and Tools

Dr. Schmidhofer provided training to senior management at Pyramid (one day) and to outpatient staff. In the TPS teacher role, he mentored Ms. Bechtold on a regular basis, providing her with hands-on training in developing A3 problem-solving tools, analyzing these A3s, and applying the TPS Rules of Use to each of the problems they identified and addressed (access, retention, clinical management).

Initial training was given to all clinicians on using the data collection instruments and understanding the various reports. Training was given to all clerical persons responsible for transmitting the data to the data center on data handling and the distribution of project reports forwarded to the program. Further training was given whenever any major changes were made to the data collection forms or project processes.

Training was included in the orientation of all staff hired during the life of the project. Training manuals were provided to each staff member.

Training (one day) was provided to the senior management team, which included Pyramid's CEO (Mr. Wolfe), chief financial officer, chief operating officer, and chief information officer. The team met monthly to discuss the project's progress and provide feedback.

The Tools Used

Project Steps

1. Determine the largest "problem" the CEO had, that is, what "kept him awake at night" with respect to the targeted programs.
2. Determine the project team and obtain CEO support to "free up" the time of one staff person to be mentored and be the lead in the project.

3. Conduct A3s regarding the "current condition" for the identified issue. Patient access was identified first because improving access would increase revenue, an outcome the CEO heartily endorsed.

4. Identify the target condition.

5. Conduct scientific tests using TPS Rules of Use, collecting data and verifying changes toward the target condition.

6. Generalize learnings and develop new protocols, manuals, and staff trainings to implement learnings.

7. Continue data collection and identify any new problems in as real time as possible for course corrections and additional learning opportunities.

Tools Used

Along with the TPS Rules of Use, instruments were developed to collect the specific information needed to measure the identified outcomes. A3s were also used.

Old Process/Plan Versus New

Treatment planning was simplified by using the individual client reports, which were produced and sent to the clinicians within one working day of receipt of the data collection instrument. The initial individual report contained much of the information necessary to complete a client treatment plan, eliminating having to page through hard copies of documents for information. Subsequent individual reports were produced each time there was a change in status for any of the outcome domains.

Clients had to be scheduled into preexisting appointment times within 10 business days. Clients were assessed and admitted for services. Evaluations were taken at the time of discharge and, by designing an end-of-treatment evaluation, of regular discharged clients only.

Clients were scheduled within 48 hours of first contact into existing and "overload" appointments. They were assessed and admitted, and data collection was begun. All clients completed updates every two or four weeks, depending on the level of care.

When there was any negative change with respect to any of the four domains (that is, progress was going in the wrong direction—for instance, the client was starting to use again), the therapist was given an individual report on that client.

Typically, these systems provide an overwhelming volume of information; staff members feel overloaded and do not use the information. With the new plan, counselors were given information that directed them to the domains in which there was a negative change. At the next client visit, the counselor was prepared to concentrate on working out whatever the specific problem was in order to continue to achieve the ultimate target of clinical excellence.

Educating Staff on New Processes

On-site training was given whenever any major changes were made to the data collection forms or additional project processes were added, including the rationale behind the changes and additions.

Biggest Obstacles

The biggest obstacle to implementing the project was getting buy-in from the staff. Initially, the staff viewed the project as merely additional paperwork. Once staff members were trained in how to use the reports and how the reports could be a useful tool in working with their clients, their attitude toward the project changed. When their suggestions were considered and often implemented, their reluctance gave way to receptivity to the change. In addition, feedback on the content and format of the reports was solicited from the staff and incorporated into the documents, when appropriate. The inclusion of their input gave staff members a sense of ownership, which further enhanced their acceptance of the project.

Easier than Expected

Adapting the reports to the voiced needs of the clinicians enabled them to see the value of the project as it applied to their particular program responsibilities. The counselors loved the process. New employees, especially, were receptive to the project because an overview of the project was presented during orientation as an integral part of the treatment process.

The Outcome

Measurable Outcomes

Measurable outcomes included the following:
- Client: employment status, criminal justice status, drug/alcohol use, and housing situation
- Program: shortening of the length of the intake process (as shown in Figures 11-1 and 11-2), client discharge status, client length of stay, and aggregate improvement in client outcomes

Time to Achieve Outcomes

Within six months of implementing "overload" appointments, monthly revenue and capacity reports revealed positive outcomes.

Time and Resources Saved

Initially, the new plan added time to the process until paperwork and other forms were streamlined and redundancy was removed. Outpatient site staff members were able to see and treat more clients with the existing staff-to-client ratio, compared with other outpatient sites that did not have this system in place. Internal process improvement reports were automatically generated for monthly meetings. Automated monthly, quarterly, and yearly reports saved hours of work.

Waste (*Muda*) Saved

Ultimately, revising all forms reduced redundancy while maintaining accreditation standards. Generated reports became part of the clinical chart and required less transfer of information to other forms.

Cost Savings

Although data for private financial reports are unavailable, increased census generated an increase in overall revenue.

Communicating Progress to Staff

Progress was shown in the monthly (aggregate progress) and individual (client progress)

FIGURE 11-1. Performance Monitoring Program: Time to First Appointment

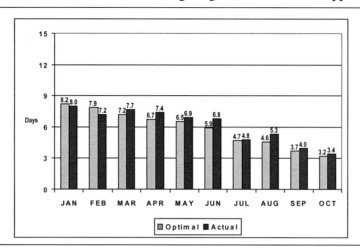

This figure shows a sample monthly aggregate report for Pittsburgh Outpatient, October 2006. Two lengths of time were calculated: (1) optimal, which is the time (in days) between the initial client contact to the program and the date offered by the program for the first appointment, and (2) actual, which is the time (in days) between the initial client contact to the program and the date the client actually reports for the first appointment. All adult and adolescent clients attending their first treatment session in a reporting month are included in the mean for that month.

Source: University of Pittsburgh, reproduced by permission.

FIGURE 11-2. Performance Monitoring Program: Time to First Treatment Session

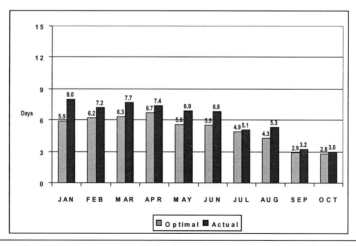

This figure shows a sample monthly aggregate report for Pittsburgh Outpatient, October 2006. Two lengths of time were calculated: (1) optimal, which is the time (in days) between the date of the first appointment attended and the date offered by the program for the first treatment session, and (2) actual, which is the time (in days) between the date of the first appointment attended and the date the client actually attended the first treatment session. All adult and adolescent clients attending their first treatment session in a reporting month are included in the mean for that month.

Source: University of Pittsburgh, reproduced by permission.

reports. Periodic meetings between the project team and program staff were held to discuss the project accomplishments.

Lessons Learned

Methodology
Interview clients and staff at both sites.

Self-Report Form
- Validity is compromised with the acuity of population (D/A vs. Dual; that is, drug and alcohol use/abuse vs. drug and alcohol use/abuse plus a mental health condition). That is, when self-report is the method of data collection, the more clinically impaired the client, the less assurance that he or she interpreted the self-report questions appropriately. Pyramid tried to identify as quickly as possible patients/clients who had reading and other problems (such as mental health issues that involved psychoses, for instance) and used volunteers to read the instruments to the patients in place of employing self-report. This tactic would have addressed literacy issues, but it still did not address the effect of a patient's thought disorder upon his or her interpretation and responses to the questions read to him or her.
- Keep data collection forms as short as possible while still collecting the necessary information.
- Pay attention to the wording of questions and the different interpretations that could be assumed.

Monthly Aggregate Reports
Using input from site staff, develop reports in the preferred format (graphs, table, text, etc.), and report only on what staff indicate will be useful in the treatment process. Review reports with staff to ensure proper interpretation and to instruct on the various methods of using the information on the reports.

Individual Reports
Delivery as close to real time as possible will enable the best use of the reports in the clinical process.

Other
More members of the organization need to understand and participate in the project so that companywide decisions will incorporate the project results and reduce additional work in other areas. Design the project to meet Joint Commission accreditation standards as well as the requirements of licensing and regulatory bodies.

Last Words
The project leader needs to be given companywide support as well as time to dedicate to the project—that means full time in order to achieve and sustain changes within the system. There is a need to identify and incorporate current use of technology and generational characteristics associated with the use of technology.

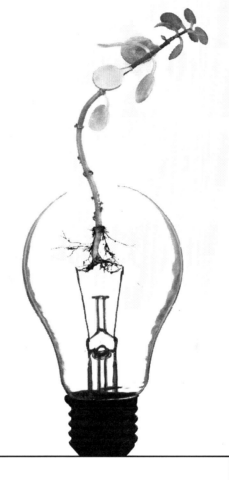

HOW TO MAKE LEAN WORK FOR YOUR ORGANIZATION:

A LOOK AT THE CANCER TREATMENT CENTERS OF AMERICA

Case At a Glance

THE ORGANIZATION:	Cancer Treatment Centers of America, Philadelphia; Zion, Illinois; and Tulsa, Oklahoma
THE LEAN PROJECT:	Lean philosophy in action throughout an entire health care organization
THE TOOLS USED:	Lean Six Sigma, A3 Performance Improvement Process (PIP), *kaizen* events, key performance indicators (KPIs)
THE OUTCOME:	Reported patient satisfaction of 99%, staff overall satisfaction of 85.1%

The Organization

Cancer Treatment Centers of America (CTCA) specializes in treating patients with complex and advanced stage (stages 3 and 4) cancers. Typical CTCA patients are those who have previously been treated elsewhere and have been dissatisfied and those with metastatic disease and as many as five comorbidities. CTCA offers state-of-the-art diagnostics and clinical care in medical, radiation, and surgical oncology in three fully functioning hospitals and one clinic. CTCA sites are located in Philadelphia; Zion, Illinois; and Tulsa, Oklahoma. Another hospital in Phoenix was expected to be completed by 2009.

CTCA was begun in 1988 and was founded on tenets of consumer experience. Its founder, Richard J. Stephenson, and his family were dissatisfied with the treatment and service given his mother, Mary Brown Stephenson, during the long months of her treatment for bladder cancer. When Ms. Stephenson lost her battle with cancer in 1982, the family members took it upon themselves to design and plan treatment centers that offered integrative and compassionate cancer care to clients and their families. Because their dedication to care arose from this experience, CTCA holds all its services to The Mother Standard®. This standard asks one simple question: "How would I want my mother treated?" Recent data reveal that 99% of patients treated at CTCA said they would "definitely bring their mother to CTCA."

Because CTCA's mission is "to never stop searching for and providing powerful and innovative therapies to heal the whole person, improve quality of life and restore hope," management recognized a perfect fit with Lean Six Sigma philosophies. Lean thinking practices have been woven throughout the organization, and all employees are required to be trained and to participate to some degree. Herb De Barba, Vice President for Lean Six Sigma at CTCA, came to CTCA in 2005 and manages the coordination and of all Lean activities.

CTCA is featured in this chapter because CTCA management has embraced Lean and Six Sigma, fully integrating Lean thinking into all levels of work. CTCA trains and educates all of its stakeholders and expects process improvement from everyone. Although this is a unique setup, this company's trailblazing work can serve as an inspiration to other organizations, step by step and commitment by commitment.

Discerning Value

Using Lean Six Sigma to improve and sustain quality, safety, and compassionate care means discerning what the patient values. For CTCA, everything begins with the customer. Of the

450 Lean projects done in 2007, 449 began with opportunities to improve patient care, whereas only 1 project was begun with an internal issue.

To determine, as opposed to assuming, what its patients value, CTCA uses surveys, focus groups, board and other meetings, a sophisticated software program, and a particularly widespread application of the Lean A3 tool. Through these methods, CTCA has learned that its patients value successful outcomes, quality of life, patient empowerment, whole-person treatment, and high value-added care delivery.

The following are just a few examples of how Lean initiatives at CTCA have improved the patient experience.

- Same-day insurance verification
- First appointment generally within two to five days
- Comprehensive, integrated treatment plan within three to five days
- Ability to start treatment immediately
- Return patient registration eliminated

Some of the specific projects that CTCA has undertaken have involved doing the following:

- Performing direct handoffs from Oncology Information Specialists (OIS) to Scheduling
- Gathering medical records prior to admission within two days
- Reducing initial registration time from 45 minutes to 7 minutes
- Keeping turnaround time (TAT) on tumor marker tests within 45 minutes
- Supplying positron-emission tomography (PET) scan results within 10 hours
- Reducing TAT for chemotherapy by 23% while increasing volume 30%
- Eliminating repeat blood draws
- Eliminating returned patient registrations
- Increasing clinic appointments (from 432 to 504)

Annual results have included the following:

- Average of $17 million per year in cost savings and capacity increases since beginning Lean Six Sigma, documented for 2005, 2006, and 2007
- 49,884 stakeholder non-value-added hours eliminated in 2007 alone
- 8,071 patient wait hours eliminated in 2007
- Medical record retrieval time reduced from five days to two days

Choosing and Using Tools

The tools each organization chooses will depend on the organization and the work to be done. The major deployment tools CTCA has used on its Lean Six Sigma journey include *kaizen* events, key performance indicators (KPIs), A3 problem-solving processes, and Six Sigma. The A3 Performance Improvement Process (PIP) is a bedrock of improvement work, and CTCA has developed its own 62-page A3 "textbook" to guide employees in its best use.

Somewhere between 5 and 7 *kaizen* events are held at each CTCA site each year; for example, a total of 18 *kaizen* events were held at CTCA in 2007. Every year, Midwestern Regional Medical Center, one of the CTCA institutions, plans a 1.5-day strategic planning session for senior and middle management. Projects are discussed, and brainstorming is used to prioritize *kaizen* projects for the upcoming year. Five *kaizen* projects are chosen, and a champion is selected for each *kaizen*. Managers voice whether they want to be a part of the team.

KPIs are regularly conducted in individual departments, and departmental KPIs and *kaizen* events are integrated. Projects created around KPIs generally establish a turnaround time objective (as the example in Figure 12-1 shows) and are tied to performance bonuses.

Examples of KPIs include projects that did the following:

- Reduced TAT on tumor marker tests from 124 to 72 minutes
- Reduced registered nurse (RN) vacancy rate from 12% to 1%
- Reduced patient transfer time from recovery to floor by 83%
- Reduced blood draw to lab TAT from 20 to 8 minutes
- Reduced check process TAT from 16 to 8 hours

Whereas *kaizen* events are considered "top down," the numerous, smaller A3 projects are "bottom up" and stakeholder driven. The A3 PIP is taught as an eight-week course with two-hour twice-weekly classes conducted with about 15 students. One hour of homework is given at each class. During each course, a departmental project is completed and topped off by a graduation ceremony.

Tools are tailored, and so are forms. For example, CTCA took seven forms used in a department and condensed them into one comprehensive computerized "document." There had been redundancy and confusion, and no single form had contained all the information that the user needed. This element was part of a much bigger project being undertaken, but this initial segment needed to be clarified so that input downstream would be just as clear. The new comprehensive Excel form offers boxes to check; checking boxes means no additional information is required, which cuts down on information overload.

There are three sets of tools taught at CTCA, and the tools used depend on the application:

1. A basic set of tools is taught to every person in the organization. The A3 performance improvement process starts with the establishment of a SMART (specific, measurable, attainable, relevant, and time limited) objective (always focus on improving the patient experience), followed by an analysis of the current state using value stream analysis, followed by root cause identification techniques and a bit of project planning in terms of implementation and control.

2. Another set of tools is taught to supervisors, managers, and directors. These tools are primarily geared toward providing a deeper understanding of how to supervise in the Lean Six Sigma environment: how to coach and mentor teams, how to keep projects on track, and how to motivate people to generate ideas.

3. The core group of Green Belts and Black Belts in each hospital also has 60 to 75 tools available to them. These experts generally lead large, complex projects and also are called in where appropriate to help someone who is having difficulty deciding which tool to use to solve a particular kind of problem.

In addition to the Lean projects held each year, there are also roughly 15 to 20 annual Six Sigma projects. They all follow the basic DMAIC process.

Advanced Education and Training

Deployment of Six Sigma mastery has increased each year at CTCA. The organization formally adopted the program, developed its own course to train belts, and continued to train its own employees. In 2005, there were no Green Belts or Black Belts. By 2006, CTCA had 14 belts;

FIGURE 12-1. Turnaround Time Between Patient Check-In and Specimen's Arrival in Lab

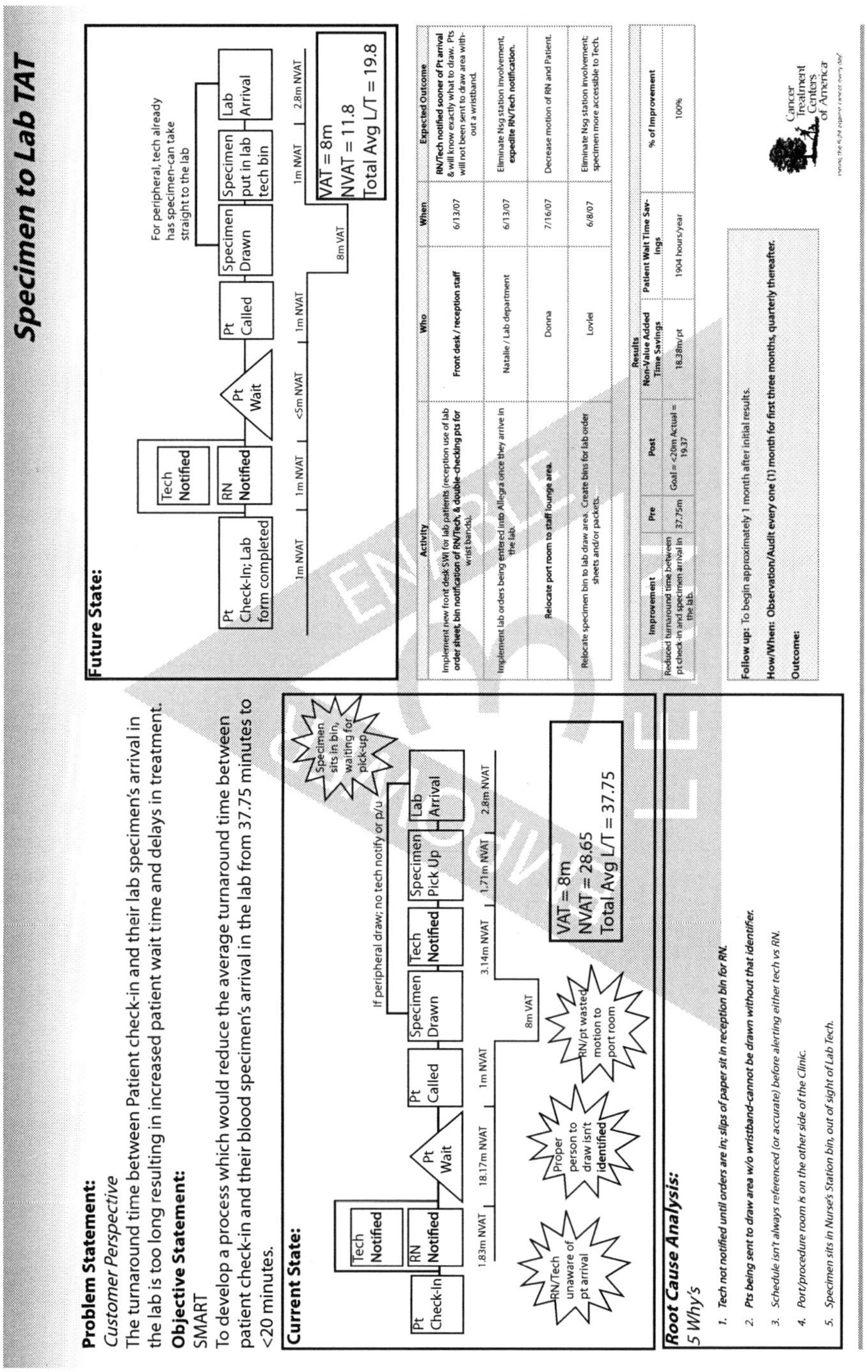

Average turnaround time (TAT) was reduced from more than 37 minutes to less than 20.

Source: Cancer Treatment Centers of America, reproduced by permission.

in 2007, there were 30 belts, and by the beginning of 2008, there were 60 belts. The Black Belt certification process has included a CEO as well as the chief medical officer (CMO).

Participation in the Green Belt education and certification process is open to all stakeholders. The objective here is to certify at least one stakeholder in each department There is also a central group comprised of approximately three people who are Black Belts, which means that 30 to 40 Green Belts in each hospital can be supported by the Black Belts of the second group. Within each hospital there are approximately 40 to 50 departments.

Participation in a certified program at the University of Michigan College of Engineering Center for Professional Development or the American Society for Quality (ASQ) provided all original Six Sigma training. Later, these consultants helped CTCA develop its own course. At each of its hospitals CTCA employs one full-time Lean Six Sigma Master Black Belt, a full-time A3 instructor, and certified Lean coaches. All management members are required to become "belted," and all stakeholders are required to become A3 certified; the A3 certifications are internally developed and delivered.

Forming and Motivating Teams

Along with patient satisfaction, staff satisfaction is important at CTCA. In the 2006 survey of CTCA stakeholders, the following results indicate how Lean is geared to promote value among these customers (stakeholders) as well:

- 85.1% expressed overall satisfaction with CTCA.
- 86% say they will recommend CTCA as a place to work to friends and associates.
- 94% reported "I like the kind of work I do at CTCA."
- 87% said their work provides them with a sense of accomplishment.
- 95% are proud of the work they do.

Although many health care companies consider Lean to be a series of *kaizen* events and teams are selected to become involved, at CTCA *kaizen* events are only one piece of the puzzle, and all employees are members of the larger team. There are two types of teams: *Kaizen* teams are one type; the other type is departmentally driven. CTCA picks multidisciplinary teams, and the champion gets the team started and reports in after every 8 to 16 hours of work to receive guidance, course correction, and support. For example, on Monday morning, leadership comes in and rallies the team, makes sure the objective is clear, and reminds team members that if roadblocks come up, support is available. In a three-to-five-day event, management can also course-correct and cheer on the troops. Typically during these events, the larger team breaks up into smaller teams to attack the implementation issues. At the end, the whole team gathers to ensure that none of the small groups' decisions will cause problems for upstream or downstream processes. Standard work is authorized, and all members leave to begin implementation. The outcome of one such event is shown in Figure 12-2.

A second type of team is the standing team that does not disperse after one concrete event (for example, the First Connections team shown in Figure 12-3, page 154). These standing teams are either departmental or interdepartmental. All front-line stakeholders who do not have someone reporting to them are being trained to use the A3 PIP. This includes everyone: physicians, nurses, aides, technicians, housekeeping, admissions personnel, and every other employee. Following the central tenets of TPS,[1] paraphrased as "All work must be con-

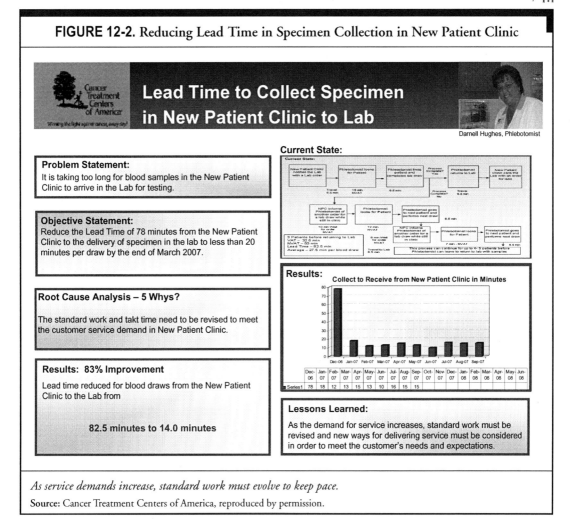

FIGURE 12-2. Reducing Lead Time in Specimen Collection in New Patient Clinic

As service demands increase, standard work must evolve to keep pace.

Source: Cancer Treatment Centers of America, reproduced by permission.

trolled, modified and constantly improved by the people who are doing the work," CTCA's expectation is that employees will continually use the A3 PIP to keep improving processes within their own domains. Among every 6 to 10 people within a department, one person will more than anyone else embrace Lean thinking. At CTCA, that person is recognized as a coach of A3 process improvement. Within that group, when the group comes up with a new idea, that coach is the one to take the group through the process, and that person gets trained above and beyond the typical employee training.

Communication in the Organization

Communication about Lean Six Sigma in any organization can be achieved through team charters, status reports, and final reports. However, informal communication can be conducted in meetings, in organizational publications, and throughout normal daily work.

Communication techniques are embedded in CTCA's strategic planning process. The leadership convenes periodically, and it also meets formally once a year. When projects are complete,

FIGURE 12-3. Storyboard for First Connections Team's Work

First Connections

Problem Statement: The Registration process is slow and cumbersome; patients are asked the same information three to four times; and there are many forms and signatures required by the patient from many different departments.

Objective Statement: The objective is to eliminate redundancy in the process, thereby reducing the time it takes for a new patient to register. The implications on patient satisfaction are vast as we should see a drastic decrease in patient complaints after implementation.

Root Cause Analysis: 5 Whys

Why is registration taking 45 minutes on average?

- Patients are backed up in the New Patient Clinic, and staff cannot keep up with the patient volume.

Why are patients backed up in the New Patient Clinic?

-Patients are displaced because asked to wait, and then arrive late.

Why are patients told to wait?

-The registration process takes a long time for patients to complete.

Why do the forms take a long time to complete?

- The forms require redundant information

- It is difficult for patients to remember all of the information required when they are feeling nervous and are in a new place.

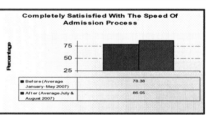

Percent of patients <u>Completely Satisfied</u> with the ease of the admission process from January to June 2007 as compared to July and August 2007. The new process was fully implemented in July 2007.

Improvements:

- Reduced lead time to register a patient from 45 minutes to less than 10 minutes.

- Replaced twelve forms with four forms. The patients complete all but two forms in the comfort of their own homes and review the I information when arrive here.

- Increased patient loyalty scores for ease of admission from 79.7% 2006 to 85.7% in 2007.

Lesson Learned:

Although First Connections takes place in the lobby area, space is still a concern. First Connections Specialists have attempted to take laptops into the atrium to complete the registration process, but due to HIPAA, they cannot discuss medical information in a public area. The First Connections team is working on creative construction solutions.

The team is still working on creative construction solutions to reduce registration process time.

Source: Cancer Treatment Centers of America, reproduced by permission.

they are communicated up to the board of directors, and at every board meeting, Mr. De Barba reports on Lean Six Sigma activities. Lean Six Sigma is also a standing agenda item at each hospital's leadership team meeting. Every A3 class graduation takes place at a board meeting where participants report on their projects. All Six Sigma graduations occur the same way. Every newsletter includes a section on Lean Six Sigma accomplishments.

REFERENCE

1. Spear S.J.: Decoding the DNA of the Toyota Production System. *Harv Bus Rev* 77(5): 97–106, Sep.–Oct. 1999.

APPENDIX:

RESOURCES

Books

Hadfield D., Holmes S.: *The Lean Healthcare Pocket Guide: Tools for the Elimination of Waste in Hospitals, Clinics and Other Healthcare Facilities.* Chelsea, MI: MCS Media, 2006.

Liker J.K.: *The Toyota Way: 14 Management Principles from the World's Greatest Manufacturer.* New York: McGraw-Hill, 2004.

Martin K., Osterling M.: *Kaizen Event Planner: Achieving Rapid Improvement in Office, Service, and Technical Environments.* New York: Productivity Press, 2007.

Womack J.P., Jones D.T.: *Lean Thinking: Banish Waste and Create Wealth in Your Corporation.* New York: Free Press, 2003.

Zak H.: *Doing More with Less: Lean Thinking and Patient Safety in Health Care.* Oakbrook Terrace, IL: Joint Commission Resources, 2006.

Zidel T.G.: *A Lean Guide to Transforming Healthcare: How to Implement Lean Principles in Hospitals, Medical Offices, Clinics and Other Healthcare Organizations.* Milwaukee: ASQ Quality Press, 2006.

Articles

Abelson D.: Lean production efforts help save $7.5M in 1 year. *Healthcare Benchmarks Qual Improv* pp. 137–138, Dec. 2005.

Advisory Board Company: *"Lean" Models at Health Care Institutions.* Original inquiry brief. Washington, DC: Marketing and Planning Leadership Council, Dec. 19, 2006.

Bahensky J.A., Roe J., Bolton R.: Lean Sigma: Will it work for healthcare? *J Healthc Inf Manag* 19:39–44, 2005.

Benneyan J.C., Lloyd R.C., Plsek P.E.: Statistical process control as a tool for research and healthcare improvement. *Qual Saf Health Care* 12:458–464, Dec. 2003.

Bertels T., Williams M., Dershin H.: Six Sigma: A powerful strategy for healthcare providers. *Aon Healthcare Alliance Health Line* Special Edition:1–5, 2001.

Bossert J.: Lean and Six Sigma: Synergy made in heaven. *Quality Progress* 36:31, Jul. 2003.

Brue G.: The elephant in the operating room. *Quality Digest,* Jun. 2005.

Bushell S., Shelest B.: Discovering Lean thinking at Progressive Healthcare. *Journal for Quality & Participation* 25(2):20–25, 2002.

Carlino A.: Five principles can encourage organizations to "think Lean." *Perform Improv Advis* pp. 94–95, Aug. 2004.

Castle L., Franzblau-Isaac E., Paulsen J.: Using Six Sigma to reduce medication errors in a home-delivery pharmacy service. *Jt Comm J Qual Patient Saf* 31(6):319–324, 2005.

Chan A.L.: Use of Six Sigma to improve pharmacist dispensing errors at an outpatient clinic. *Am J Med Qual* 19(3):128–131, 2004.

de Koning H., et al.: Lean Six Sigma in healthcare. *J Healthc Qual* 28(2):4–11, 2006.

5S: A Lean method to cut the clutter. *OR Manager* 23(3):15, 2007.

Furman C.: Implementing a patient safety alert system. *Nurs Econ* 23:42–45, Jan.–Feb, 2005.

Furman C., Caplan R.: Applying the Toyota Production System: Using a patient safety alert system to reduce error. *Jt Comm J Qual Patient Saf* 33:376–386, Jul. 2007.

Hagland M.: Six Sigma: It's real, it's data-driven, and it's here. *Health Care Strateg Manage* 23:1, 13–16, 2005.

APPENDIX:

RESOURCES

Hill D.: Physician strives to create lean, clean health care machine: Studies of manufacturing processes may one day help make your practice more efficient. *Physician Exec* 27:62–65, Sep.–Oct. 2001.

Kolodziej J.H.: The Lean team. *Mich Health Hosp* 37:24–26, Jan.–Feb. 2001.

Lazarus I.R., Andell J.: Providers, payers and IT suppliers learn it pays to get "Lean." *Managed Healthcare Executive*, Feb. 2006.

Lean thinking: Eliminating waste and adding value to OR processes. *OR Manager* 23(3):1, 2007.

Long J.C.: Healthcare Lean. *Mich Health Hosp* 39:54–55, Jul.–Aug. 2003.

Martin K.: Lean thinking. Slide presentation given at Joint Commission Resources Ambulatory Care Conference, Chicago, Oct. 1, 2007.

Martin K.: On Lean enterprise and its potential healthcare applications. *J Healthc Qual* 25:2, 43, Sep.–Oct. 2003.

McAuliffe J.: Practicing "wasteology" in the OR. *OR Manager* 23(3):10–15, 2007.

Nelson-Peterson D.L., Leppa C.J.: Creating an environment for caring using Lean principles of the Virginia Mason Production System. *J Nurs Adm* 37:287–294, Jun. 2007.

Pittsburgh Regional Healthcare Initiative puts new spin on improving healthcare quality. *Qual Lett Healthc Lead* 14:2–11, 21, Nov. 2002.

Printezis A., Gopalakrishnan M.: Current pulse: Can a production system reduce medical errors in health care? *Qual Manag Health Care* 16:226–238, 2007.

Rozgus A.: Using the sixth sense: By implementing the Six Sigma approach, companies can move ahead of the pack. *Concrete Producer,* Aug. 1, 2003.

Scalise D.: Six Sigma: The quest for quality. *Hosp Health Netw* 75(12):41–46, 2001.

Six practices of the Lean operating room. *OR Manager* 23(3):10, 2007.

Sobek D., Jimmerson C.: A3 reports: Tool for process improvement. Paper presented at the Industrial Engineering Research Conference Proceedings, Houston, May 16–18, 2004.

Spath P.: Is your organization thinking Lean? *Hosp Peer Rev* 29:99–100, Jul. 2004.

Spear S.J.: Decoding the DNA of the Toyota Production System. *Harv Bus Rev* 77(5):97–106, Sep.– Oct. 1999.

Spear S.J.: Fixing health care from the inside, today. *Harv Bus Rev* 83:78–91, 158, 2005.

Spear S.J.: Learning to lead at Toyota. *Harv Bus Rev* 82(5):78–86, 151, 2004.

Sunyog M.: Lean Management and Six-Sigma yield big gains in hospital's immediate response laboratory. Quality improvement techniques save more than $400,000. *Clin Leadersh Manag Rev* 18:255–258, Sep.–Oct. 2004.

Thompson D.N., Wolf G.A., Spear S.J.: Driving improvement in patient care: Lessons from Toyota. *J Nurs Adm* 33:585–595, Nov. 2003.

Weber D.O.: Toyota-style management drives Virginia Mason. *Physician Exec* 32(1):12–17, 2006.

Womack J.P., Jones D.T.: Beyond Toyota: How to root out waste and pursue perfection. *Harv Bus Rev* 74:140–158, Sep.–Oct. 1996.

Zidel T.G.: A Lean toolbox: Using Lean principles and techniques in healthcare. *J Healthc Qual* 28: W1-7–W1-15, 2006.

Online Documents

Breakthrough Management Group International: *Lean Six Sigma Case Studies.* http://www.bmgi.com/success_stories/casestudies.aspx (accessed Mar. 21, 2008).

Duhig J.M.: *Six Sigma and Lean Production.* Intermountain Healthcare. http://intermountainhealthcare.org/xp/public/institute/library/faculty16.xml (accessed Mar. 20, 2008).

Grout J.: *Mistake-Proofing the Design of Health Care Processes.* AHRQ Publication No. 07-0020. Agency for Healthcare Research and Quality (AHRQ; prepared under an IPA with Berry College), May 2007. http://www.ahrq.gov/qual/mistakeproof/mistakeproofing.pdf (accessed May 7, 2008).

Healthcare Performance Partners: *Labor and Delivery Realize Annualized Savings of $94K, & Increases Bedside Time by 33%.* http://leanhealthcareperformance.com/healthcaredocuments/HPPCaseStudyL&D.pdf (accessed Oct. 21, 2008).

Healthcare Performance Partners: *Nursing Team Redesigns Floor and Eliminates Waste.* http://leanhealthcareperformance.com/healthcaredocuments/HPPCaseStudyNursingRedesign.pdf (accessed Mar. 18, 2008).

Lowstuter B.: *Mistake-Proofing Through Source Checks, Self Checks and Successive Checks.* Healthcare Performance Partners, Mar. 12, 2008. http://leanhealthcareexchange.com/?p=102 (accessed Sep. 22, 2008).

Manivannan S.: Error proofing. *MRO Today,* Dec. 2006/Jan. 2007. http://www.mrotoday.com/mro/archives/Mfg%20excellence/ErrorProofingDJ07.htm (accessed Mar. 21, 2008).

Manos A., Sattler M., Alukal G.: Make healthcare Lean. *Quality Progress,* Jul. 2006. http://www.asq.org/qualityprogress/past-issues/index.html?fromYYYY=2006&fromMM=07&index=1 (accessed Mar. 21, 2008).

Norbut M.: Driving for efficiency: Saving time and money while boosting quality. Virginia Mason Medical Center in Seattle got its philosophy of a Lean production system from an unlikely source: Toyota. *Am Med News* 48, Oct. 3, 2005. http://www.ama-assn.org/amednews/2005/10/03/bisa1003.htm (accessed May 9, 2008).

Simon K.: *Poka Yoke Mistake Proofing.* i Six Sigma Healthcare, 2008. http://healthcare.isixsigma.com/library/content/c020128a.asp (accessed Mar. 21, 2008).

Smith B.: Lean and Six Sigma: A one-two punch. *Quality Progress* 36(4), Apr. 2003. http://www.asq.org/qualityprogress/past-issues/index.html?fromYYYY=2003&fromMM=04&index=1 (accessed Mar. 21, 2008).

Web Sites

Advance for Health Information Executives (Advanceweb.com)
http://health-care-it.advanceweb.com/editorial/content/editorial.aspx?cc=92458

American Society for Quality
http://www.asq.org/learn-about-quality/lean/overview/read-more.html

Gemba Research
http://www.gemba.com/Healthcare.cfm

"Going Lean in Health Care" (white paper on Lean)
http://www.ihi.org/IHI/Results/WhitePapers/GoingLeaninHealthCare.htm.

Institute for Healthcare Improvement (general health care)
http://www.ihi.org/IHI/Topics/Improvement/ImprovementMethods/Changes/

Iowa Healthcare Collaborative
http://www.ihconline.org/toolkits/leaninhealthcare.cfm

i Six Sigma Healthcare
http://healthcare.isixsigma.com/

Lean Advisors, Inc.
http://www.leanadvisors.com/Lean/Healthcare/health_casestudies.cfm

Lean Enterprise Institute
http://www.lean.org/

Lean Healthcare Exchange
http://www.leanhealthcareexchange.com/

Lean Healthcare Services
http://www.leanhealthcareservices.com/lean-healthcare-explained.htm

Lean Healthcare West
http://leanhealthcarewest.com/lean_healthcare_links.html#

Superfactory.com
http://www.superfactory.com/lean-healthcare.html

GLOSSARY

affinity diagram: Used after brainstorming, the affinity diagram helps organize ideas so that connections between issues are logical.

A3 problem solving: A reporting system used to document efforts to solve problems concisely and to promote understanding of processes by seeking consensus of all parties that will be affected by proposed changes.

autonomation: A term coined by Shigeo Shingo, Toyota Production System pioneer, which roughly translates as preautomation and is equivalent to "stopping the line."

batching and queuing: In contrast to single piece flow, batching is the mass production of large lots or parts so that waiting time (queuing) exists before sending the batch to the next operation in the process. In Lean, batching and queuing are eliminated or kept to an absolute minimum.

brainstorming: The process of capturing people's ideas, without censoring or editing them, and organizing those thoughts around common themes.

brownfield: Space that is currently occupied and that will be redesigned.

cause-and-effect diagram: *See* fishbone diagram.

continuous flow: The ability to provide a service or item when requested, without delay or rework. After perfecting the ability to "see" value and waste, creating continuous flow is the next practical step. (*See also* just-in-time, *takt* time.)

control chart: The visual representation of tracking progress over time.

cycle time: The time it takes to complete one cycle of an operation. *See also* process time.

DMAIC: The acronym for the five phases of Six Sigma implementation: defining, measuring, analyzing, improving, controlling.

document tagging: A technique for accurately capturing the amount of time it takes for a chart, document, or patient to travel through a process, area, or value stream. The document or item is "tagged" with dates and times.

downstream: The patient, provider, and/or process that requires an upstream service and is paying for it.

FIFO: Signifies "first in, first out" lanes in a value stream map or standard work flow. This tool ensures that the oldest work upstream (first in) is the first to be processed downstream (first out).

fishbone diagram (also called cause-and-effect diagram or *Ishikawa* diagram): The visual representation to clearly display the various factors affecting a process. This can be a structured approach to root cause analysis. The diagram identifies the inputs or potential causes of a single output or effect. In a hospital, for example, work can be divided into categories: responding to instructions (orders), using supplies and medications (materials), using equipment (machinery), providing the needed care or service in accordance with established procedures (methods), and the environment itself (environment). Those categories can be listed on a fishbone diagram as the different branches from which mistakes can arise. To complete the branches, brainstorm primary causes, ask together why they are occurring, analyze the causes, and prioritize and identify the likely root causes.

5S: A process to ensure that work areas are systematically kept clean and organized, ensuring employee safety and providing the foundation on which to build a Lean culture.

5 Whys tool: A means of getting at root causes, this technique involves continuing to ask "Why?" until a root cause is identified. Use of structured questions, including specific inquiries, facilitates this technique's success. This tool is especially helpful in combination with a 5S event.

flow: The continuous movement and progressive achievement of tasks along a value stream. To accomplish better flow, eliminate constraints or barriers that impede movement of the patient or product from one operation or process to the next.

flowchart: A diagram showing a step-by-step process, depicted as symbols connected by lines. Basic flowcharting symbols include Supplies, Database, Sequence, *Kanban,* Information, Inventory, and Improvement. Each has its own icon. Note: A flowchart is similar to a value stream map but is bigger and much more detailed. Process flowcharting is used to analyze and improve a process, but value stream mapping cuts across processes, departmental and functional boundaries, and existing performance-measurement systems.

gemba: The production floor, the front line, the source of learning in the workplace.

greenfield: Currently unoccupied space requiring a design from the start rather than a redesign.

histogram: The visual representation displaying the spread and shape of the data distribution (the distribution away from the mean).

Ishikawa **diagram:** *See* fishbone diagram.

just-in-time (JIT): The process of supplying the customer with the precisely needed product or service, in the right amount, at the requested time, whenever it is needed, every time it is needed.

kaizen: From the Japanese words *kai* and *zen;* may be translated as "improvement." A *kaizen* event is an intensive workshop to tackle a focused problem and to set a path for continuous incremental improvement over a sustained period.

kanban: The Japanese word meaning "signal card." A *kanban* is most commonly a visual card or other indicator, attached to supplies or equipment, that serves as a means of communicating to an upstream process precisely what is required at the specified time. *Kanban* is a method of just-in-time production that uses standard containers or lot sizes with a single card attached to each.

lead time: The time a customer must wait to receive a product or service after requesting it. Lead time is the time that elapses when work has been made available until it has been completed and passed along to the next step in the process. This is also known as throughput, turnaround, and elapsed time.

metric: A specific number that is used to measure before and after initiatives.

muda: The Japanese word for "waste." Designates any activity that uses resources but provides no value for the customer.

non-value-added: Any steps in a process that do not directly contribute to the customer's (patient's, other stakeholder's) experience of value. The customer may or may not be willing to pay for a non-value-added product or service, but some non-value-added steps are required despite this distinction.

Pareto chart: Also called the Pareto diagram or Pareto analysis, this is a bar graph that displays the activities being studied in order from largest to smallest. The Pareto chart is useful in analyzing the frequency of problems or causes in a process, when wanting to focus on the most significant problems or concerns, when analyzing broad causes by looking at their specific components, and when communicating about data.

Paynter chart: Based on the Pareto principle (*see* Pareto chart), this chart focuses on the areas of priority but graphically displays data by subgroups.

physical layout: The arrangement of work spaces, supplies, and pathways within the workplace. A good physical layout can eliminate a lot of waste, including motion and transportation, and can optimize the flow of people, work, and information. *See also* continuous flow.

poka-yoke: The Japanese term for error proofing.

process: A set of actions or steps, each of which is to be undertaken appropriately and accomplished properly in sequence at the appropriate time in order to create value for the patient. Key processes, or value streams, are those that support core products or services. In health care, they may include a doctor's office visit, a lab visit, a hospital stay, or a visit to the emergency department.

process mapping: The visual representation of a sequence of tasks consisting of people, work duties, and transactions that occur for the design and delivery of a product or service.

process time: The time it takes to do the work without interruptions. This is also known as touch time, work time, or cycle time. (*See also* cycle time.)

pull system: A system driven by the needs of the downstream lines. By specifying value, identifying the value stream, and creating flow, Lean thinking allows pull to take place. In a Lean health care system, patients pull the product along rather than having the marketplace push it onto them on the organization's timetable. Pull offers more flexibility and accommodates changes in customer demand.

push system: Work that is driven by the output of the preceding lines. This is work that is pushed along regardless of need or request. It involves providing a service or product in anticipation of a need and is often associated with high inventory and the risk of errors or higher error rates. In Lean, push must be eliminated and replaced with a pull system to facilitate flow.

rapid process improvement (RPI) event: Another name for a *kaizen* event. Also called rapid improvement project, rapid improvement workshop, rapid improvement event.

red tag: A label used in the 5S process to identify items that are not needed or that are placed in the wrong area.

root cause: The origin or source of a problem.

root cause analysis (RCA): Process used to identify root causes or core origins of problems. RCA is best used in assessing rare events, such as wrong-surgery or egregious medication errors, as opposed to common patient safety problems, such as hospital-acquired infection or contrast-induced nephropathy. The four steps of RCA include data collection, causal factor charting, root cause identification, and generating and then implementing recommendations.

run charts: Data graphs run over time. These charts are important, frequently used performance improvement tools. The benefits of run charts include depicting how well a process is performing, displaying a pattern of data that can be observed during changes, and providing direction about the value of particular changes.

scatter diagram: A visual representation of data used to study the possible relationship between one variable and another. The scatter diagram graphs pairs of numerical data; one variable is charted on each axis. If the variables are correlated, the points will fall along a line or curve; the better the correlation, the tighter the points will adhere to the line.

sensei: The Japanese word for a personal teacher who has mastery of a body of knowledge.

Six Sigma: The measure of variation that achieves 3.4 defects per million opportunities, or 99.99966% acceptability.

Six Sigma DFSS (Design for Six Sigma): A methodology that addresses building a new product, service, or process so that it is error free and meets customer requirements with Six Sigma precision. DFSS uses the DMADV approach (Define, Measure, Analyze, Design, Verify).

standard work: The precise description of relevant information to document the best practice of producing a work unit or providing a service. Standard work aims to minimize non-value-added steps in processes, provide high quality, and reduce variation.

takt **time:** The rate of customer demand. The term is derived from the German word for "beat," "pulse," or "measure," which the Japanese began to use in the 1930s when they were learning aircraft production from German aerospace engineers. *Takt* time informs you of customer demand and often is used to synchronize the production rate to meet customer demand. In general, when cycle time equals *takt* time, you will be capable of producing to customer demand.

team charter: A document detailing the team's mission and proposed outcomes to ensure strategic alignment.

upstream: The provider of a service or work unit required downstream from a customer.

value-added time: The time element that the customer is willing to pay for. This time should be void of waste.

value stream: A sequence of processes that are connected by a common customer, product, or service request.

visual control: The visual indicators used to ensure a process produces what is expected and if not, what must happen. Visual controls communicate information nonverbally to alert people to the work process and its schedule. When something is off track or not on time, this becomes obvious.

waste: Anything that adds cost or time without adding value. The Japanese word for "waste" is *muda*.

Work-Out: A "fast" approach to process improvement in which a trained facilitator leads a group of selected key individuals who "work out" a solution to a problem using a variety of tools, such as affinity diagrams, fishbone diagrams, and action plans.

INDEX

Note: A letter *t* following a page number indicates a table, a letter *f* indicates a figure.